T0140385

Advances in Experimental Medicine and Biology

Neuroscience and Respiration

Volume 1070

Subseries Editor
Mieczyslaw Pokorski

More information about this series at http://www.springer.com/series/13457

Mieczyslaw Pokorski

Editor

Progress in Medical Research

 Springer

Editor
Mieczyslaw Pokorski
Opole Medical School
Opole, Poland

ISSN 0065-2598   ISSN 2214-8019   (electronic)
Advances in Experimental Medicine and Biology
ISBN 978-3-030-07826-3   ISBN 978-3-319-89665-6   (eBook)
https://doi.org/10.1007/978-3-319-89665-6

This Springer imprint is published by the registered company Springer Nature Switzerland AG.
The registered company address is: Gewerbestrasse 11, 6330 Cham, Switzerland

Preface

The book series Neuroscience and Respiration presents contributions by expert researchers and clinicians in the multidisciplinary areas of medical research and clinical practice. Particular attention is focused on pulmonary disorders as the respiratory tract is up front at the first line of defense for organisms against pathogens and environmental or other sources of toxic or disease-causing effects. The articles provide timely overviews of contentious issues or recent advances in the diagnosis, classification, and treatment of the entire range of diseases and disorders, both acute and chronic. The texts are thought as a merger of basic and clinical research dealing with biomedicine at both the molecular and functional levels and with the interactive relationship between respiration and other neurobiological systems, such as cardiovascular function, immunogenicity, endocrinology and humoral regulation, and the mind-to-body connection. The authors focus on modern diagnostic techniques and leading-edge therapeutic concepts, methodologies, and innovative treatments. The action and pharmacology of existing drugs and the development and evaluation of new agents are the heady area of research. Practical, data-driven options to manage patients are considered. New research is presented regarding older drugs, performed from a modern perspective or from a different pharmacotherapeutic angle. The introduction of new drugs and treatment approaches in both adults and children is also discussed.

Body functions, including lung ventilation and its regulation, are ultimately driven by the brain. However, neuropsychological aspects of disorders are still mostly a matter of conjecture. After decades of misunderstanding and neglect, emotions have been rediscovered as a powerful modifier or even the probable cause of various somatic disorders. Today, the link between stress and health is undeniable. Scientists accept a powerful psychological connection that can directly affect our quality of life and health span. Psychological approaches, which can decrease stress, can play a major role in disease therapy.

Neuromolecular and carcinogenetic aspects relating to gene polymorphism and epigenesis, involving both heritable changes in the nucleotide sequence and functionally relevant changes to the genome that do not involve a change in the nucleotide sequence, leading to disorders, are also tackled.

Clinical advances stemming from molecular and biochemical research are but possible if research findings are translated into diagnostic tools, therapeutic procedures, and education, effectively reaching physicians and patients. All this cannot be achieved without a multidisciplinary, collaborative, bench-to-bedside approach involving both researchers and clinicians. The role of science in shaping medical knowledge and transforming it into practical care is undeniable.

Concerning respiratory disorders, their societal and economic burden has been on the rise worldwide, leading to disabilities and shortening of life-span. COPD alone causes more than three million deaths globally each year. Concerted efforts are required to improve this situation, and part of those efforts are gaining insights into the underlying mechanisms of disease and staying abreast with the latest developments in diagnosis and treatment regimens. It is hoped that the articles published in this series will assume a leading position as a source of information on interdisciplinary medical research advancements, addressing the needs of medical professionals and allied health-care workers, and become a source of reference and inspiration for future research ideas.

I would like to express my deep gratitude to Paul Roos, and Cynthia Kroonen of Springer Nature NL for their genuine interest in making this scientific endeavor come through and in the expert management of the production of this novel book series.

Mieczyslaw Pokorski

Contents

Advs Exp. Medicine, Biology - Neuroscience and Respiration (2018) 39: 1–7
DOI 10.1007/5584_2018_159
© Springer International Publishing AG 2018
Published online: 15 Feb 2018

Baker's Asthma: Is the Ratio of Rye Flour-Specific IgE to Total IgE More Suitable to Predict the Outcome of Challenge Test Than Specific IgE Alone

V. van Kampen, I. Sander, R. Merget, T. Brüning, and M. Raulf

Abstract

Usually the diagnosis of baker's asthma is based on specific inhalation challenge with flours. To a certain extent the concentration of specific IgE to flour predicts the outcome of challenge test in bakers. The aim of this study was to evaluate whether the ratio of specific IgE (sIgE) to total IgE (tIgE) improves challenge test prediction in comparison to sIgE alone. Ninety-five bakers with work-related respiratory symptoms were challenged with rye flour. Total IgE, sIgE, and the sIgE/tIgE ratio were determined. Receiver operator characteristic (ROC) plots including the area under the curve (AUC) were calculated using the challenge test as gold-standard. Total IgE and sIgE concentrations, and their ratio were significantly higher in bakers with a positive challenge test than in those with a negative one ($p < 0.0001$, $p < 0.0001$, and $p = 0.023$, respectively). In ROC analysis, AUC was 0.83 for sIgE alone, 0.79 for tIgE, and 0.64 for the ratio. At optimal cut-offs, tIgE, sIgE, and the ratio reached a positive predicted value (PPV) of 95%, 84% and 77%, respectively. In conclusion, calculating the ratio of rye flour-sIgE to tIgE failed to improve the challenge test prediction in our study group.

Keywords

Allergy · Baker's asthma · Immunoglobulin E · Inhalation challenge · Occupational allergy · Rye flour

1 Introduction

Baker's asthma is one of the most frequent forms of occupational immunoglobulin E (IgE)-mediated allergy. In 2014, 64% of 584 confirmed cases of occupational asthma in Germany were caused by bakery-derived allergens, especially wheat and rye flour (Deutsche Gesetzliche Unfallversicherung (DGUV 2015)). In general, but especially within the scope of compensation claims, the specific inhalation challenge with suspected occupational allergens is considered the gold standard for the diagnosis of occupational asthma (Vandenplas et al. 2017; Muñoz et al. 2014). Since the specific challenge test is cumbersome, has a potential for severe adverse effects, and should only be performed at specialized centers, alternative methods to diagnose flour allergy are of great value.

The general association between the clinical responsiveness to allergens and the results of

V. van Kampen (✉), I. Sander, R. Merget, T. Brüning, and M. Raulf
Institute for Prevention and Occupational Medicine of the German Social Accident Insurance, Institute of the Ruhr University (IPA), Bochum, Germany
e-mail: kampen@ipa-dguv.de

specific IgE (sIgE) tests is well known. Especially for food allergens (Buslau et al. 2014; Soares-Weiser et al. 2014) but also for ubiquitous inhalation allergens (Douglas et al. 2007; Fernández et al. 2007) it was shown that the sIgE determination can be useful to predict the result of an oral or bronchial challenge. Concerning inhalative occupational allergens, it has been demonstrated for latex and flours that high levels of sIgE have a high positive predictive value (PPV) for occupational asthma (Vandenplas et al. 2016; van Kampen et al. 2008).

The level of tIgE, which is the sum of all sIgE, is mostly elevated in case of allergic sensitization. Thus, a given sIgE level might be associated with the total IgE level, which is a potential indicator of atopy (Ahmad Al Obaidi et al. 2008). For this reason, in a couple of studies the ratio of allergen-sIgE to tIgE has been used to predict the outcome of food challenge. While some authors have found a benefit using the sIgE/tIgE ratio compared with sIgE alone (Horimukai et al. 2015; Gupta et al. 2014), others have reported the opposite (Grabenhenrich et al. 2016; Mehl et al. 2005).

Therefore, this study seeks to define the diagnostic utility of the ratio of rye flour-sIgE to tIgE in the diagnosis of baker's asthma. Due to the fact that tIgE levels are mostly elevated in atopic subjects, data were additionally stratified according to atopy status.

2   Methods

2.1   Study Design

The study was approved by the Ethic Committee of the Ruhr University Bochum in Germany, and all study participants gave informed written consent. Ninety-five bakers (mean age of 40 ± 13 years, range of 19–76 years, 79% of males) with work-related symptoms of asthma and/or rhinitis were included in the study. All subjects answered a questionnaire and underwent a clinical examination. In addition, tIgE and rye flour-sIgE determinations, as well as a challenge test with rye flour were performed. All the bakers

were examined within the scope of claims for compensation due to occupational asthma.

2.2   IgE Determination

Total IgE and rye flour-sIgE were measured by ImmunoCAP (Thermo Fisher Scientific; Phadia AB, Uppsala, Sweden) according to the manufacturer's recommendations. sIgE values ≥ 0.35 kU/L were considered positive, and for lower values a value of two-thirds of the detection limit (0.23 kU/L) was assigned. For data analysis, sIgE values greater than 100 kU/L were replaced by 110 kU/L. The measuring range of tIgE for an undiluted sample is 2–5000 kU/L.

2.3   Rye Flour Challenge Test

Bronchial challenge tests were performed with nebulized aqueous rye flour solutions in 24 cases, with native rye flour simulating the situation at the workplace in 63 cases, and nasal challenges in 8 cases. A nasal challenge test was considered positive if nasal symptoms were followed by a decrease of nasal flow by at least 40% from baseline. Allergen-induced airway responsiveness was measured by body plethysmography. A positive test result was assumed if specific airway resistance (sRaw) doubled and increased to ≥ 2.0 kPa·s, or if the fall in forced expired volume in one second (FEV_1) was $\geq 20\%$.

2.4   Skin Prick Test

A panel of common inhalation allergens, including grass pollen, birch pollen, house dust mite (*Dermatophagoides pteronyssinus*), and cat dander (Allergopharma; Reinbek, Germany) were tested (double estimation). Atopy was defined as a mean wheal diameter ≥ 3 mm to at least one of these aeroallergens. Histamine (10 mg/mL) and saline were used as positive and negative controls, respectively.

2.5 Statistical Analysis

Concentrations of rye flour-sIgE, tIgE, and the sIgE/tIgE ratio in bakers with a positive or a negative challenge test were compared with the Mann-Whitney U test. Receiver operating characteristic (ROC) plots are one possible graphical presentation for describing and comparing diagnostic tests. The area under the curve (AUC) is a global measure of the diagnostic performance of a test and ranges from 0 to 1. ROC plots were constructed using the outcome of challenge test as gold-standard. AUCs were calculated to compare the results for rye flour-sIgE alone, tIgE, and the sIgE/tIgE ratio. For further evaluation, optimal cut-off levels that lead to the maximum Youden Index (sensitivity + specificity − 1) were calculated. Using these optimal cut-offs, sensitivity, specificity, positive predictive value (PPV), and negative predictive value (NPV) were calculated. Correlation between variables was assesses with Spearman's correlation coefficient (r). Comparisons were considered significant at p < 0.05. Calculations were performed using a commercial GraphPad Prism v7.03 statistical package (GraphPad Software; La Jolla, CA).

3 Results

Sixty-three (66%) of the 95 bakers challenged with rye flour showed a positive test result. The rye flour-sIgE concentration in all bakers ranged from <0.35 to >100 kU/L (median of 1.31 kU/L), tIgE from 2.3 to 4161 kU/L (median: 98.5 kU/L), and the sIgE/tIgE ratio was between 0.003 and 19.1% (median of 2.3%). The tIgE and sIgE concentrations and the sIgE/tIgE ratio were significantly higher in bakers with a positive challenge test than in those with a negative one (Fig. 1). Overall, sIgE correlated strongly with the tIgE concentration of (r = 0.688; p < 0.0001).

The AUC of the ROC plots was 0.83 for rye flour-sIgE, 0.79 for tIgE, and 0.64 for the sIgE/tIgE ratio (Fig. 2). Sensitivity, specificity, PPV, and NPV at the optimal cut-off levels (sIgE >2.4 kU/L, tIgE

>51.8 kU/L, and sIgE/tIgE >1.8%) are shown in Table 1. The maximum Youden Index was similar for sIgE and tIgE (0.556 and 0.561, respectively), while it was only half as high for sIgE/tIgE (0.276). Whereas tIgE >51.8 kU/L was a sensitive predictor of the challenge result (sensitivity 87%, NPV 75%), rye flour-sIgE was an excellent and specific predictor of a positive challenge result (rye flour-sIgE >2.4 kU/L: specificity 94%, PPV 95%).

According to skin prick testing with ubiquitous allergens, 44 bakers showed a positive reaction to at least one of the allergens and were defined as atopics. The remaining 51 bakers were defined to be non-atopic. While the concentrations of rye flour-sIgE and tIgE were significantly higher in atopic bakers, no significant difference were obtained for the sIgE/tIgE ratio (Table 2). ROC analysis stratified according to atopy is shown in Fig. 3. While in the group of non-atopic bakers, the AUC for rye flour-sIgE and tIgE were approximately 50% higher than that for their ratio, in the group of atopic bakers all three AUC were in a similar range.

4 Discussion

According to Kleine-Tebbe (2012), IgE results of symptomatic subjects should generally be checked for sIgE to tIgE ratio. In the present study we evaluated the ratio of ray flour-sIgE to tIgE in a group of 95 bakers with work-related allergic symptoms to determine whether the ratio could be superior to sIgE alone in the diagnosis of baker's asthma. As shown by the AUC values of ROC curves, the sIgE/tIgE ratio failed to improve the prediction of the outcome of a challenge test with rye flour. This finding was also true for the stratified analysis of both atopic and non-atopic bakers.

Gupta et al. (2014) have reported that the sIgE/tIgE ratio is significantly more accurate than sIgE alone in predicting the oral challenge outcome in children with suspected food allergy (AUC 0.69 vs. 0.55; p = 0.03). In contrast, in another study this ratio has not been as efficacious as sIgE alone for the diagnosis of symptomatic food allergy in children and therefore judged of no real benefit

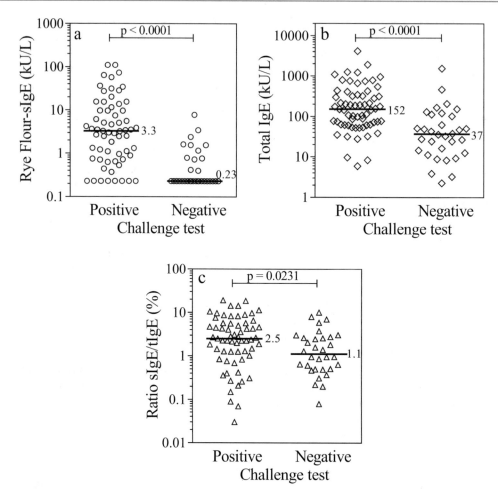

Fig. 1 Comparison of concentrations of rye flour-sIgE (**a**), total IgE (tIgE) (**b**), and the ratio of sIgE/tIgE (**c**) between bakers with a positive challenge test (n = 63) and those with a negative challenge test (n = 32) to rye flour. Horizontal lines in each panel represent medians

(Mehl et al. 2005). In the latter study, differences in the AUCs of ROC curves depended on the allergen. The respective AUCs were 0.79 (ratio) *vs.* 0.76 (sIgE alone) for cow's milk, 0.86 *vs.* 0.88 for hen's egg, 0.66 *vs.* 0.66 for wheat, and 0.58 *vs.* 0.64 for soy. For the allergens above outlined, the maximal sIgE/tIgE ratio amounted to 91.2% (median: 0.3%) for cow's milk, 69.4% (median: 1.7%) for hen's egg, 70.7% (median: 0%) for wheat, and 15.0% (median: 0%) for soy. From this ranking, one could suppose that the utility of the ratio in predicting the outcome of a challenge test increases with the height of the ratio. This could be an explanation for the low diagnostic

performance of the rye flour-sIgE/tIgE ratio in the present study (AUC 0.64 (ratio) *vs.* 0.83 (sIgE alone)) because the maximum ratio was only 19.1% (median: 2.3%).

However, the afore-mentioned studies were performed in children suffering from food allergy, while this study concerns adult bakers with respiratory work-related symptoms subjected to inhalation exposure to rye flour. In this connection, it is noteworthy that the level of tIgE is age-dependent. In a study including 603 children, aged between 1 and 9 years, with suspected food allergy a significant positive correlation between the total serum IgE level and patients' age has been observed (Horimukai et al. 2015). On the other hand, a large longitudinal study spanning more than a decade in 6371 young and middle-aged adults (20–44 years) has shown that tIgE decreases in elderly subjects while there was no significant change in the prevalence of sensitization to at least one environmental allergen over the study period (Jarvis et al. 2005).

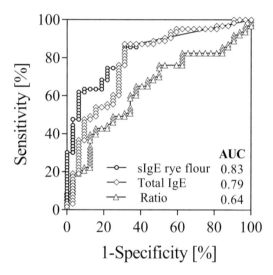

Fig. 2 Receiver operating curves (ROC) for rye flour-sIgE, total IgE (tIgE), and the ratio of sIgE/tIgE of 95 symptomatic bakers with the challenge test as gold-standard. The area under curve (AUC) is a global measure for the test's accuracy that ranges between 0 and 1

One study enrolling 171 adult β-lactam allergic patients and 122 controls shows that the sIgE/tIgE ratio could improve the diagnosis of β-lactam allergy (Vultaggio et al. 2015). Due to the fact that the authors compared the diagnostic performance of sIgE to at least one β-lactam hapten with the ratio that was calculated by making the sum of sIgE to all four measured β-lactam haptens divided by tIgE, a comparison of the usefulness of sIgE alone and the ratio, like we did it in the present study, was not possible. In addition, a skin prick test with β-lactams rather than a challenge test was used as gold-standard in that previous study. Nonetheless, the authors hypothesize that a high sIgE/tIgE ratio increases the probability to have two neighboring sIgE molecules on the basophil/mast cell surface membrane, which could lead to an easier cross-linking of the high-affinity IgE receptor with cell activation that follows.

An additional finding of the present study was that the level of tIgE resulted in a much better prediction of the outcome of a challenge test with rye flour than the ratio of rye flour-sIgE/tIgE (AUC: 0.79 vs. 0.64), although it could be expected due to a high correlation between rye flour-sIgE and tIgE. To the best of our knowledge, so far no study has evaluated the predictive value of tIgE for the outcome of an inhalation allergen challenge, but it is known that high levels of tIgE strongly increase the probability of sensitization,

Table 1 Sensitivity, specificity, positive (PPV) and negative predictive value (NPV) of rye flour-specific IgE (sIgE), total IgE (tIgE), and the sIgE/tIgE ratio, based on the gold-standard challenge test

	Cut-off	Maximum Youden Index	Sensitivity (%)	Specificity (%)	PPV (%)	NPV (%)
sIgE (kU/L)	> 2.4	0.556	61.9	93.8	95.1	57.0
tIgE (kU/L)	> 51.8	0.561	87.3	68.8	84.4	74.5
sIgE/tIgE (%)	> 1.8	0.276	65.1	62.5	77.1	49.1

Evaluation was performed using the cut-offs which were obtained by the maximum Youden Index

Table 2 Comparison of concentrations of rye flour-specific IgE (sIgE), total IgE (tIgE), and the sIgE/tIgE ratio between atopic and non-atopic bakers

	Atopics (n = 44) median (range)	Non-atopics (n = 51) median (range)	p-value
sIgE (kU/L)	2.75 (0.23–110)	0.72 (0.23–57.1)	0.007
tIgE (kU/L)	134 (5.91–4161)	72 (2.28–1536)	0.036
sIgE/tIgE (%)	2.25 (0.03–18.60)	2.04 (0.07–19.10)	0.401

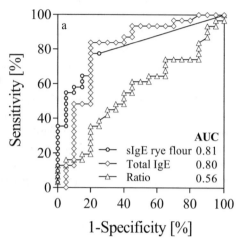

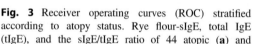

Fig. 3 Receiver operating curves (ROC) stratified according to atopy status. Rye flour-sIgE, total IgE (tIgE), and the sIgE/tIgE ratio of 44 atopic (**a**) and 51 non-atopic (**b**) bakers were evaluated with the challenge test as gold-standard; AUC, area under curve

especially in young individuals (Kerkhof et al. 2003) and also are predictive of asthma (Ahmad Al Obaidi et al. 2008). However, tIgE should not be used as definite evidence for an allergic disease since its level is influenced by numerous factors, especially age (Chang et al. 2015).

In conclusion, rye flour-sIgE is superior to the ratio of rye flour-sIgE to tIgE, independent from atopy status, as a diagnostic test to distinguish between bakers with and without occupational asthma to rye flour. Overall, the accuracy of the ratio, as a diagnostic test, was not satisfactory in the present study. Whereas tIgE resulted in a high negative predicted value and was more suitable to rule out a positive result of a challenge test, it could be confirmed that a high concentration of rye flour-sIgE in the sera of bakers suffering from work-related symptoms is a good predictor of a positive challenge result. Thus, specific challenges with flours may be omitted in strongly sensitized bakers (van Kampen et al. 2008).

Acknowledgements We gratefully acknowledge Ursula Meurer who performed the IgE measurements.

Conflicts of Interest The authors declare no conflicts of interest in relation to this article.

References

Ahmad Al Obaidi AH, Mohamed Al Samarai AG, Yahya Al Samarai AK, Al Janabi JM (2008) The predictive value of IgE as biomarker in asthma. J Asthma 45:654–663

Buslau A, Voss S, Herrmann E, Schubert R, Zielen S, Schulze J (2014) Can we predict allergen-induced asthma in patients with allergic rhinitis? Clin Exp Allergy 44:1494–1502

Chang M-L, Cui C, Liu Y-H, Pei L-C, Shao B (2015) Analysis of total immunoglobulin E and specific immunoglobulin E of 3,721 patients with allergic disease. Biomed Rep 3:573–577

DGUV (2015) Deutsche Gesetzliche Unfallversicherung Statistics, Sankt Augustin (in German)

Douglas TA, Kusel M, Pascoe EM, Loh RK, Holt PG, Sly PD (2007) Predictors of response to bronchial allergen challenge in 5- to 6-year-old atopic children. Allergy 62:401–407

Fernández C, Cárdenas R, Martín D, Garcimartín M, Romero S, de La Cámara AG, Vives R (2007) Analysis of skin testing and serum-specific immunoglobulin E to predict airway reactivity to cat allergens. Clin Exp Allergy 37:391–399

Grabenhenrich L, Lange L, Härtl M, Kalb B, Ziegert M, Finger A, Harandi N, Schlags R, Gappa M, Puzzo L, Stephan V, Heigele T, Büsing S, Ott H, Niggemann B, Beyer K (2016) The component-specific to total IgE ratios do not improve peanut and hazelnut allergy diagnoses. J Allergy Clin Immunol 137:1751–1760

Gupta RS, Lau CH, Hamilton RG, Donnell A, Newhall KK (2014) Predicting outcomes of oral food challenges by using the allergen-specific IgE-total IgE ratio. J Allergy Clin Immunol Pract 2:300–305

Horimukai K, Hayashi K, Tsumura Y, Nomura I, Narita M, Ohya Y, Saito H, Matsumoto K (2015) Total serum IgE level influences oral food challenge tests for IgE-mediated food allergies. Allergy 70:334–337

Jarvis D, Luczynska C, Chinn S, Potts J, Sunyer J, Janson C, Svanes C, Künzli N, Leynaert B, Heinrich J, Kerkhof M, Ackermann-Liebrich U, Antó JM, Cerveri I, de Marco R, Gislason T, Neukirch F, Vermeire P, Wjst M, Burney P (2005) Change in prevalence of IgE sensitization and mean total IgE with age and cohort. J Allergy Clin Immunol 116:675–682

Kerkhof M, Dubois AE, Postma DS, Schouten JP, de Monchy JG (2003) Role and interpretation of total serum IgE measurements in the diagnosis of allergic airway disease in adults. Allergy 58:905–911

Kleine-Tebbe J (2012) Old questions and novel clues. Complexity of IgE repertoires. Clin Exp Allergy 42:1142–1145

Mehl A, Verstege A, Staden U, Kulig M, Nocon M, Beyer K, Niggemann B (2005) Utility of the ratio of food-specific IgE/total IgE in predicting symptomatic food allergy in children. Allergy 60:1034–1039

Muñoz X, Cruz MJ, Bustamante V, Lopez-Campos JL, Barreiro E (2014) Work-related asthma. Diagnosis and prognosis of immunological occupational asthma and work-exacerbated asthma. J Investig Allergol Clin Immunol 24:396–405

Soares-Weiser K, Takwoingi Y, Panesar SS, Muraro A, Werfel T, Hoffmann-Sommergruber K, Roberts G, Halken S, Poulsen L, van Ree R, Vlieg-Boerstra BJ, Sheikh A (2014) The diagnosis of food allergy. A systematic review and meta-analysis. Allergy 69:76–86

van Kampen V, Rabstein S, Sander I, Merget R, Brüning T, Broding HC, Keller C, Müsken H, Overlack A, Schultze-Werninghaus G, Walusiak J, Raulf-Heimsoth M (2008) Prediction of challenge test results by flour-specific IgE and skin prick test in symptomatic bakers. Allergy 63:897–902

Vandenplas O, Froidure A, Meurer U, Rihs H-P, Rifflart C, Soetaert S, Jamart J, Pilette C, Raulf M (2016) The role of allergen components for the diagnosis of latex-induced occupational asthma. Allergy 71:840–849

Vandenplas O, Suojalehto H, Cullinan P (2017) Diagnosing occupational asthma. Clin Exp Allergy 47:6–18

Vultaggio A, Virgili G, Gaeta F, Romano A, Maggi E, Matucci A (2015) High serum β-lactams specific/total IgE ratio is associated with immediate reactions to β-lactams antibiotics. PLoS One 10(4):e0121857

Advs Exp. Medicine, Biology - Neuroscience and Respiration (2018) 39: 9–18
DOI 10.1007/5584_2018_156
© Springer International Publishing AG 2018
Published online: 20 Feb 2018

SERPINA1 Gene Variants in Granulomatosis with Polyangiitis

Malgorzata Hadzik-Blaszczyk, Aneta Zdral,
Tadeusz M. Zielonka, Ada Rozy, Renata Krupa,
Andrzej Falkowski, Kazimierz A. Wardyn,
Joanna Chorostowska-Wynimko, and Katarzyna Zycinska

Abstract

Alpha-1 antitrypsin (A1AT) deficiency is one of the most common genetic disorders in Caucasian population. There is a link between granulomatosis with polyangiitis (GPA) and most frequent variants of SERPINA1 gene encoding severe alpha-1 antitripsin deficiency. However, the potential effect of Pi*Z, Pi*S as well as other SERPINA1 variants on clinical course of vasculitis are not well understood. The aim of the study was to analyze the potential effect of A1AT protein phenotype representing the SERPINA1 gene variants on the clinical course of GPA. The study group consisted of 64 subjects with GPA, stratified according to the disease severity: patients in active phase (group I, n = 12), patients during remission on treatment (group II, n = 40) or untreated (group III, n = 12). Normal Pi*MM SERPINA1 genotype was detected by means of real-time polymerase chain reaction (PCR) or direct sequencing in 59 patients, Pi*MZ genotype in 2, and Pi*IM, Pi*MS or Pi*SZ in 1 patient respectively. The patients with abnormal Pi*Z, Pi*S, or Pi*I allele constituted 17% in group I, 5% in group II, and 8% in group III. The serum content of A1AT and high sensitivity C-reactive protein (hsCRP) assessed by nephelometry did not differ between the groups. Interestingly, the mean serum antiPR3-antibodies level detected by Elisa method was significantly greater in the GPA patients with Pi*Z, Pi*S, or Pi*I SERPINA1 variants than in the Pi*MM homozygotes. In summary, heterozygous Pi*MZ, Pi*MS, and Pi*SZ genotype was detected in 7.8% of total group of GPA patients, and in 10.5% of those with lung lesions. The abnormal alleles of Pi*S and Pi*Z may affect the clinical course of the disease.

Keywords

Alpha-1 antitrypsin deficiency · AntiPR3-antibodies · C reactive protein · Disease activity · Genotyping · Granulomatosis with polyangiitis · Phenotyping

M. Hadzik-Blaszczyk, T. M. Zielonka (✉), R. Krupa, A. Falkowski, K. A. Wardyn, and K. Zycinska
Department of Family Medicine, Internal and Metabolic Diseases, Warsaw Medical University, Warsaw, Poland
e-mail: tadeusz.zielonka@wum.edu.pl

A. Zdral, A. Rozy, and J. Chorostowska-Wynimko
Department of Genetics and Clinical Immunology, National Institute of Tuberculosis and Lung Diseases, Warsaw, Poland

1   Introduction

Granulomatosis with polyangiitis (GPA) is one of the primary systemic vasculitides, an autoimmune disease of unknown etiology. There is a number of known factors important for its pathogenesis including infectious (viral or bacterial), environmental (exposure to silicon), iatrogenic (drugs) (Millet et al. 2014; Stassen et al. 2009; Barnett et al. 1999) and of genetic origin such as *HLA DP, SERPINA1* variants or *PRTN3* gene (encoding a serine proteinase-3) (Lyons et al. 2012). GPA is characterized by a complex immunological response to necrosis of small vessels resulting from inflammation. Antineutrophil cytoplasmic antibodies (cANCA or PR3-ANCA against serine proteinase-3 and pANCA against myeloperoxidase) play an important role in GPA pathogenesis (Xiao et al. 2016). It has been suggested that vascular necrosis in the course of GPA might, in part, result from the enhanced activity of proteolytic enzymes. GPA is a chronic disease characterized by periods of exacerbations and remissions, necessitating therapy with potent anti-inflammatory and immunosuppressive drugs (Yates et al. 2016). Likewise, the clinical presentation including the involvement of a specific organ and the activity of immune responses are diverse. Thus, the disease activity is monitored not only by clinical status but also by markers of inflammation including C-reactive protein (CRP) level (Kronbichler et al. 2016).

Alpha-1 antitrypsin (A1AT), encoded by the *SERPINA1* gene, is one of the major plasma and tissue inhibitors of serine proteases such as neutrophil elastase, proteinase-3, myeloperoxidase, cathepsin G, and trypsin (McKinney et al. 2014). Alpha-1 antitrypsin deficiency is one of the most common genetic disorder in the Caucasian population. It results in systemic imbalance of proteolytic enzymes and their inhibitors, leading to excessive activation of the inflammatory process. Nearly a 130 variants of the *SERPINA1* gene have been described as being responsible for specific quantitative or qualitative disorders of A1AT (Lara et al. 2014). Severe deficiency is characterized by a low content of A1AT in the serum and lungs and a significantly higher risk of early onset emphysema and COPD, bronchial asthma, liver disease, but also GPA (Stockley and Turner 2014). The S and Z *SERPINA1* mutations are the most common cause of A1AT deficiency in homozygotes Pi*ZZ, Pi*SS or heterozygotes Pi*SZ (Popławska et al. 2013; Janciauskiene et al. 2011). The number of carriers of MZ or MS phenotype is estimated at about 116 million and the number of patients with A1AT deficiency (Pi*SZ, SS, ZZ genotype) is approximately 3.4 million in the world (Chorostowska-Wynimko et al. 2016). The potential effects of Pi*Z, Pi*S and other rare *SERPINA1* variants on the activity and clinical course of vasculitis are not well understood. Patients with abnormal A1AT phenotypes have a significantly higher vasculitis activity as well as anti-proteinase 3 (anti-PR3) antibody content (Pervakova et al. 2016). It has been suggested that Pi*Z heterozygosity is a marker of poor prognosis (Elzouki et al. 1994). Therefore, the aim of the present study was to evaluate the effect of A1AT protein phenotype representing the *SERPINA1* gene variants on the clinical course of GPA.

2   Methods

2.1   Study Population

The study protocol was approved by a local Bioethical Committee of the National Institute of Tuberculosis and Lung Diseases in Warsaw, Poland. The study included 64 patients with GPA referred to the Clinical Department of Family Medicine, Internal Medicine and Metabolic Diseases of Warsaw University Czerniakowski Hospital. All patients had GPA diagnosed according to the American College of Rheumatology criteria (Lutalo and D'Cruz 2014) including clinical symptoms, histopathological confirmation, and serum ANCA positivity. In all cases, chest X-ray and computed tomography scans were performed. The patients were enrolled prospectively between 2014 and 2015. Based on

Table 1 Patient characteristics

	Group I	Group II	Group III
	Induction phase or relapse therapy	Remission phase, with treatment	Remission phase, without treatment
Patients (n)	12	40	12
Women	6	24	11
Men	6	16	1
Mean age (year)	46.1 ± 13.1	57.1 ± 12.3	51.5 ± 15.6
Cyclophosphamide and corticosteroids	10	15	0
Cyclophosphamide and corticosteroids and plasma exchange	1	0	0
Corticosteroids in monotherapy	1	17	0
Corticosteroids and azathioprine	0	6	0
Corticosteroids and mycophenolate mofetil	0	2	0
Without treatment	0	0	12

the clinical evidence and the current method of treatment: three groups of patients was identified (Table 1). The first group consisted of 12 patients on induction therapy or with GPA relapse on intensive treatment. Induction therapy consisted of parenteral pulsed cyclophosphamide 15 mg/kg (three pulses every 2 weeks, then 3–6 pulses every 3 weeks). Alternatively, the therapy consisted of oral cyclophosphamide at a dose of 2 mg/kg/day for 3–6 months and intravenous methylprednisolone (500–1000 mg for 3 consecutive days) or oral methylprednisolone at a dose of 1 mg/kg/day, as prednisone equivalent, for at least 1 month, with subsequent individual gradual dose tapering to a maintenance dose on remission. In patients with rapidly progressive renal failure and diffuse alveolar hemorrhage, induction therapy was supplemented with plasma exchange. In patients with serious adverse side effects due to immunosuppressants, glucocorticoids were used in monotherapy. A second group consisted of of 40 patients in the remission phase on maintenance treatment with prednisone – 7.5–12.5 mg/day, cyclophosphamide – 1.5 mg/kg/day, azathioprine 2 mg/kg/day, and mycophenolate mofetil – 2 g/day. A third group consisted of 12 untreated patients in remission. The mean age of all patients (41 female and 23 male) was 53.6 ± 13.8 years.

2.2 Clinical Biochemistry

The content of serum high-sensitivity C-reactive protein (hsCRP) was assessed using a commercially available nephelometry, with the normal range set at ≤ 0.80 mg/dL. The evaluation of A1AT in the serum and in dried blood spots, normal range of 88–180 mg/mL, was assessed using the nephelometric system (Immage 800; Beckman Coulter, Clare, Ireland). Anti-proteinase 3 (anti-PR3) antibodies, negative <20 RU, were assessed in the serum using a QuantaLyser ELISA testsystem (Inova Diagnostics; San Diego, CA).

2.3 Phenotyping and Genotyping

The A1AT phenotype assessment as performed in the serum and in in dried blood spots using isoelectrofocusing on agarose gel with the Hydrasys electrophoresis (Sebia; Lisses, France) that utilizes immunofixation and a specific A1AT antibody.

Genomic DNA was extracted from dried blood spots using a commercial Extract-N-Amp Blood PCR kit (Sigma-Aldrich; St. Lois, MO). Genetic material present in the eluate was directly used for A1AT genotyping without a need of DNA purification from blood. The identification of the two most common mutations of the A1AT gene (Z, S) was performed in a single reaction by the real-time PCR in the LightCycler 480 II instrument (Roche Diagnostics; Basel, Switzerland) using hydrolyzing probes coupled with fluorescent dyes (VIC or FAM) complementary to the mutant variants (Pi*S or Pi*Z). Primer and probe sequences as well as PCR reaction conditions have been previously described in detail (Chorostowska-Wynimko et al. 2015).

Diagnosis of rare A1AT variants was established by direct sequencing. Sequence analysis of A1AT exons 2–4 was performed in a 16-capillary 3130xl Genetic Analyzer (Applied Biosystems; Foster City, CA).

2.4   Statistical Evaluation

Data were presented as means ±SD. The equality of variances was verified with Levene's test and normality of data distribution with Shapiro-Wilk's test. Inter-group differences in the A1AT content were evaluated with a *t*-test. Differences in the hsCRP and antiPR3-antibody contents in GPA patients with normal and abnormal alleles were evaluated with the non-parametrical Mann-Whitney U test. A p-value <0.05 defined statistical significance for the differences above outlined. A correlation between A1AT and hsCRP contents was evaluated with the Spearman R test, with the level of significance set at $p \leq 0.01$.

3   Results

3.1   *SERPINA1* Gen Variants

The major *SERPINA1* gene variants were identified by serum A1AT protein phenotyping and confirmed with genotyping or sequencing. The Pi*MM genotype was identified in 92% patients, whereas Pi*Z and Pi*S alleles in 5% and 3% subjects, respectively. There were two

Pi*MZ, one Pi*MS, one Pi*IM, and one Pi*SZ heterozygotes among the patients. No homozygosity was identified for the deficient alleles.

3.2   A1AT, hsCRP, and AntiPR3-Antibodies

The mean serum A1AT content was 156.6 ± 36.1 mg/dL (min-max 69.8–267.0 mg/dL) in the whole group. The content below a lower limit of normal of A1AT of 88 mg/mL was found in three patients, while above the upper limit of normal of 180 mg/mL in 13 patients. The mean hsCRP was 0.85 ± 1.62 mg/dL (min-max 0.03–11.40 mg/dL). The values above the upper limit of normal 0.80 mg/dL were observed in 16 patients. There was a significant correlation between the serum hsCRP and A1AT ($r = 0.20$, $p < 0.05$) (Fig. 1). The mean contents of hsCRP and A1AT in patients in remission (on or without treatment) tended to be lower than those in patients in the active stage of the disease, but the difference failed to reach the statistically significance ($p < 0.13$ and $p < 0.09$, respectively) (Table 2).

There was a significant difference in the mean serum A1AT between patients with Pi*MM phenotype and those with alleles Z, S, or I ($p = 0.00002$) (Table 3). The mean antiPR3-antibodies in the GPA patients with non-M alleles (Pi*Z, Pi*S, or Pi*I) was higher than that in Pi*MM homozygotes ($p = 0.052$) (Fig. 2).

3.3   Clinical Findings and Course of GPA Patients

Frequency of lung lesions in the patients stratified according to the A1AT phenotype was determined (Table 4). The Pi*MM genotype accounted for 92% patients. In seven patients (11%), a chronic respiratory disease coexisted with GPA, mainly COPD ($n = 3$), previously treated tuberculosis ($n = 2$), isolated bronchiectasis ($n = 2$), and bronchiectasis coexisting with COPD ($n = 1$) or with tuberculosis ($n = 1$). All patients diagnosed with chronic respiratory

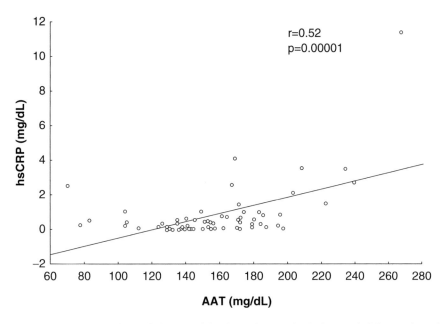

Fig. 1 Correlation between serum level of high-sensitivity C-reactive protein (hsCRP) and alpha-1 antitrypsin (A1AT)

Table 2 Content of A1AT and hsCRP in groups of GPA patients

Group	A1AT (mg/dL)	hsCRP (mg/dL)
I	172.0 ± 50.0	1.56 ± 3.90
II	156.0 ± 31.0	0.64 ± 0.77
III	145.0 ± 30.0	0.86 ± 1.27
Total	156.6 ± 36.1	0.85 ± 1.62

Data are means ±SD. Group I – disease induction phase or relapse therapy; Group II – remission phase, with treatment; Group III – remission phase, with no treatment

disease presented with the Pi*MM genotype. The patients with abnormal Pi*Z, Pi*S, or Pi*I allele constituted 17% in Group I, 5% in Group II, and 8% in Group III. However, differences between groups were insignificant. The most common abnormalities in chest CT scan images were nodules and fibrosis, which was present in 71% of patients. Bronchiectasis was found in 32%, ground-glass opacity in 20%, and emphysema in 21% of patient. Cavities were the least likely observed changes as they were present in 11.5% of patients. Pi*Z and Pi*S alleles were present in 11% of patients with pulmonary changes versus about 8% in the whole GPA group. All patients with Pi*MZ, Pi*MS, or Pi*SZ phenotype (n = 4) presented with pulmonary abnormalities in chest

CT scan images. Nodules and fibrosis were most frequently observed (n = 3), while in one Pi*MS patient nodules appeared as isolated anomaly. Three patients presented with the ground-glass opacities, co-existing with emphysema and bronchiectasis. There were no lung lesions found in the Pi*IM patient.

Only had five patients (8%) a history of smoking in the total GPA group. All ex-smokers presented with a normal Pi*MM genotype. In the group of ex-smokers, nodules with fibrosis were observed in three patients, while nodules with emphysema in one patient.

No clinically significant liver functional abnormality was observed in the GPA patients. Serum bilirubin was in the normal range in all, while modestly increased alanine aminotransferase (ALT), up to 103 U/L (normal range: 10–41 U/L), was detected in six patients (9.4%). The patient with a rare Pi*IM phenotype presented slightly elevated ALT (51 U/L). Increased activity of aspartate aminotransferase (AST) of 41 U/L (normal range: 10–31 U/L) was observed in one patient (1.6%) who was infected with hepatitis B. An ultrasound examination demonstrated hepatomegaly in four patients

Table 3 Comparison of laboratory indices in GPA patients with normal and abnormal A1AT phenotypes

	GPA patients with Phenotype Pi*MM (n = 59)	GPA patients with phenotypes Pi*Z, S, or I (n = 5)	p
A1AT level (mg/dL)	162.0 ± 32.4	93.6 ± 18.6	0.00002
hsCRP content (mg/dL)	0.85 ± 1.67	0.89 ± 1.00	0.470
AntyPR3-antibodies (RU/mL)	19.0 ± 8.0	23.3 ± 15.0	0.052

Data are means ±SD

Fig. 2 Anti-proteinase 3 (anti-PR3) antibodies in granulomatosis with polyangiitis (GPA) patients with normal MM alleles and with abnormal Pi*Z, Pi*S, and Pi*I alleles

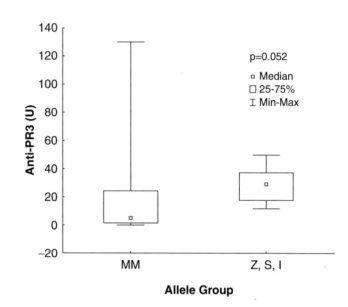

Table 4 A1AT phenotype in the whole group of GPA patients and in patients with lung lesions

Phenotype	Patients (n)	Patients with lung lesions (n)
MM	59	34
MZ	2	2
MS	1	1
SZ	1	1
IM	1	0

(6.3%), all having a Pi*MM genotype and a normal ALT serum activity. Hepatic steatosis was observed in 19 patients (29.7%), including one Pi*MZ and 18 Pi*MM patients. ALT was increased in four of those patients, all with the Pi*MM genotype. Co-existence of hepatomegaly and hepatic steatosis was observed in two patients (3.1%); the Pi*MZ and Pi*MM patients. There were liver cysts observed in ultrasound examination in two patients (3.1%) with the Pi*MM genotype, neither accompanied by changes in liver function.

The individual course of GPA in patients with deficient *SERPINA1* variants was evaluated (Table 5). In two Pi*MZ patients, acute onset of the disease, multi-organ involvement of upper and lower respiratory tracts, eyes and ears, but no renal involvement were observed. A similar pattern was present in the Pi*IM patient in whom GPA started sharply at the age of 35, with the involvement of upper and lower respiratory tracts. The disease was a dynamic process with many relapses. In turn, patient with the Pi*SZ genotype presented mild disease with the involvement of the upper and lower respiratory tracts and kidney, and a history of successful treatment followed by remission with no further need for medication.

Table 5 Description of clinical and radiological findings in GPA patients with pathological A1AT phenotypes

No.	1	2	3	4	5
Age (year)	69	40	39	28	49
Gender	Female	Male	Male	Female	Female
A1AT phenotype	MZ	MZ	MS	SZ	IM
A1AT level (mg/mL)	104.0	77.3	122.0	69.8	105.0
hsCRP level (mg/mL)	1.07	0.288	0.107	2.55	0.416
antyPR3-antibodies (RU/mL)	37.0	17.7	49.8	11.8	29.7
ALT activity (U/L)	18	27	16	17	51
Group	II	I	II	III	I
Disease duration (year)	4	1	6	4	14
Lung involement	Ground glass, fibrosis	Nodules, fibrosis, emphysema	Nodules	Nodules, fibrosis, bronchiectasis	None
Creatinine (mg/dL)	1.1	0.7	1.5	0.8	0.7
Microhematuria	–	–	+		–
Ultrasound hepatomegaly/steatosis	+	–	–	–	–
Relapses	Yes	Onset	Yes	No	Yes

hsCRP reference value of ≤ 0.8 mg/dL, A1AT reference value of 88–180 mg/mL, ALT reference value of 10–31 U/L, antyPR3-antibodies reference values: negative <20 RU/mL, moderately positive 21–30 RU/mL, positive >30 RU/mL

The only Pi*MS patient had a history of acute GPA onset at the age of 33, treated with cyclophosphamide pulses and prednisone, subsequently in remission while on continuous therapy with prednisone and azathioprine.

4   Discussion

The *SERPINA1* heterozygosity does not result in the A1AT serum content below 11 μM; the level defining severe A1AT deficiency. However, it has been suggested that A1AT deficiency alleles, including heterozygosity, predisposes to certain systemic disorders, including GPA (Fregonese and Stolk 2008). In the present study, deficient *SERPINA1* variants were detected in 7.8% of patients with GPA (Pi*Z in 4.7%, Pi*S, or Pi*I in 3.2%). Such a small number of patients precludes any unequivocal conclusion regarding the potential role of Pi*S and Pi*Z in clinical course of GPA. Previous studies have reported the Pi*Z allele in 5–27% of GPA patients (Pervakova et al. 2016; Mahr et al. 2010; Baslund et al. 1996). Other studies suggest a stronger association between GPA and Pi*Z than GPA and Pi*S variant (Chorostowska-Wynimko et al.

2013; Hersh et al. 2004; Elzouki et al. 1994). Yet Morris et al. (2011) have found no relationship between GPA and the Pi*S deficiency allele, while Mahr et al. (2010) have found a similar frequency of both Pi*Z (7.4%) and Pi*S (11.5%) alleles in a considerable group of 433 Caucasian GPA patients. These discrepancies might result from the differences in the populations studied. Mahr et al. (2010) investigated the prevalence of A1AT deficiency alleles in the US, while Morris et al. (2011) in Western European. Ferrarotti et al. (2012), who evaluated the results of the Swiss Cohort Study on Air Pollution and Lung Diseases in Adults (SAPALDIA), have demonstrated the presence of a normal Pi*MM phenotype in 89%, Pi*MS in 7.5%, Pi*MZ in 2.4%, and Pi*ZZ in 0.02% of GPA patients. The incidence of rare and new alleles was 0.7%. Variations in the incidence of deficiency alleles and irregular distribution of alleles in GPA patients clearly depends on the population studied (Blanco et al. 2006). Geographical differences in the prevalence of main A1AT gene variants, with a higher frequency of Pi*Z in Northern Europe and Pi*S in the Iberian Peninsula, are recognized.

In the present study, there were more patients carrying Pi*Z than Pi*S allele among the GPA patients. That coincides with a significantly higher prevalence of Pi*Z over Pi*S allele, as demonstrated in the screening of newborns for A1AT deficiency alleles performed in the Mazovian region of Poland (Chorostowska-Wynimko et al. 2012). With respect to the rare SERPINA1 variants, one subject with Pi*I heterozygosity was identified in the present study. The Pi*I allele is characterized by Arg39Cys transition in the amino acid sequence, which causes the aggregation of a defective A1AT protein in hepatocytes and results in both quantitative and qualitative A1AT deficiency. The Pi*I variant contributes to the development of emphysema and liver disease (Popławska et al. 2013). Yet, in the context of heterozygosity, Pi*I presence might not result in any clinically significant organ abnormalities.

Lung lesions were found in 59% of GPA patients in the present study, which is in line with other reports estimating the lung involvement on 50–90% of patients. It has been implied that disease duration determines the severity of lung lesions (Morris et al. 2011). The most common abnormalities in CT images include: pulmonary nodules (70–80%) or pulmonary infiltrates and cavitating infiltrates (50%) (Gómez-Gómez et al. 2014; Wiatr and Gawryluk 2013). The incidence of nodules observed in the present study was similar (71%), yet cavities were present in only 10.5% of patients, which was due possibly to a high number of subjects in remission. The percentage of deficiency alleles tended to be greater in the GPA patients with respiratory involvement (10.5%) when compared with the entire group of patients (7.8%). All patients with deficiency alleles presented abnormal CT images. There is a strong link between the severe A1AT deficiency and lung emphysema as well as smoking. It is estimated that 1–2% of COPD cases is due to A1AT deficiency (Chorostowska-Wynimko et al. 2016). There were no current smokers among the SERPINA1 heterozygotes (Pi*MZ, Pi*IM, Pi*MS, and Pi*SZ) in the present study. However, nodules and fibrosis were observed more frequent than emphysema, bronchiectasis, and ground glass opacities in CT images. Due to a low number of patients with non-Pi*M SERPINA1 variants, it was not possible to decisively determine clinical implications of these genetic abnormalities. However, in the group of patients with abnormal the Pi*Z, Pi*S, or Pi*I genotype an interesting pattern of more frequent GPA relapses was observed. Interestingly, this pattern was accompanied by a greater antiPR3-antibody level in the subgroup of GPA SERPINA1 heterozygotes. Pervakova et al. (2016) have observed the same pattern in a group of 38 GPA patients, including 7 with SERPINA1 deficiency variants. Since A1AT modulates PR3 activity, it might have a considerable influence on the severity of the inflammatory process in GPA. Indeed, we demonstrate that the mean A1AT serum level in GPA patients with atypical A1AT phenotypes was significantly lower than that in patients with Pi*MM phenotype. These results implicate an important protective role of A1AT in systemic vasculitis and therefore a tendency for more severe GPA in individuals with abnormal A1AT activity due to deficient SERPINA1 variants. This notion is further supported by mortality data coming from the long-term follow up. Elzouki et al. (1994) have reported a 38% death rate in the Pi*Z heterozygote GPA group versus 17% in Pi*MMs, which suggests that Pi*Z heterozygosity may indeed be linked to poor prognosis. Our results suggest that a similar picture might be true for other abnormal SERPINA1 variants. However, further studies are needed, involving appreciably larger groups of patients with a prolonged follow-up to be able to draw final conclusions on the link between heterozygosity and prognosis in GPA.

The A1AT is an acute-phase protein. Therefore, any acute or chronic inflammatory process affects its serum content. For that reason, it is recommended to perform parallel assessment of A1AT and CRP serum contents, when screening for A1AT deficiency, particularly in subjects with a confirmed or suspected chronic inflammatory disorder such as GPA. The present findings confirm a direct correlation between A1AT and CRP serum levels in the population of GPA patient. There also was a trend toward a higher A1AT content in patients with active GPA compared with disease remission. Hence, the practical conclusion coming from our observations would be that A1AT serum content in any GPA patient should be considered normal only if paralleled by CRP value within normal limits. Otherwise, diagnostic caution is advised, particularly when screening for A1AT deficiency. The possibility of a false negative result should be taken into consideration and, eventually, the A1AT protein phenotyping or *SERPINA1* genotyping performed. The decision to discontinue further diagnostics due seemingly to normal A1AT content may result in failure to detect up to 80% of deficiency variants, mainly in the Pi*MZ heterozygotes (Ottaviani et al. 2011). Alternatively, a direct sample referral for A1AT phenotyping or genotyping should be considered in GPA patients.

In the present study, the lowest A1AT content (69.8 mg/dL) was observed in a GPA patient with Pi*SZ, untreated due to remission, with augmented hsCRP (2.6 mg/dL) and abnormal chest CT scan images (nodules, interstitial fibrosis, and bronchiectasis). Another case of a low A1AT value (77.3 mg/dL), with normal serum hsCRP (0.3 mg/dL), was observed in a patient with the Pi*MZ phenotype during relapse treated with cyclophosphamide and glucocorticoids, and with similarly abnormal CT images. In yet another patient with the Pi*MZ phenotype, a normal level of A1AT (104.0 mg/dL), with elevated hsCRP (1.1 mg/dL), was found during maintenance therapy with azathioprine and glucocorticoids. Pulmonary interstitial fibrosis and ground glass appearance was seen in chest CT scan images in this case.

We conclude that Pi*MZ, Pi*MS, and Pi*SZ phenotypes are found in 7.8% of all patients with granulomatosis with polyangiitis and in 10.5% of those with the involvement of the lungs. The abnormal alleles of Pi*S or Pi*Z may affect the clinical course of the disease. The level of antiPR3-antibodies is greater in the patients with Pi*S, Pi*Z, and *SERPINA1* variants than in the Pi*MM homozygotes.

Acknowledgements This study was performed as part of the scientific project 'Dissemination and optimization of alpha-1-antitrypsin deficiency diagnostic algorithm in patients with chronic lung diseases' (Theme 5/4), of the National Institute of Tuberculosis and Lung Diseases in Warsaw, Poland. The authors are deeply indebted to the Polish Foundation for Patients with Alpha-1 Antitrypsin Deficiency for support of this study. Authors also wish to thank Ms. Natalie Czaicki for the English revision of the manuscript.

Conflicts of Interest The authors declare no conflicts of interest in relation to this article.

References

Barnett T, Sekosan M, Khurshid A (1999) Wegener's granulomatosis and α1-antitrypsin-deficiency emphysema: proteinase-related diseases. Chest 116:253–255

Baslund B, Szpirt W, Eriksson S, Elzouki AN, Wiik A, Wieslander J, Petersen J (1996) Complexes between proteinase 3, alpha-1-antitrypsin and proteinase 3 antineutrophil cytoplasm autoantibodies: a comparison between alpha 1-antitrypsin PiZ allele carriers and non-carriers with Wegener's granulomatosis. Eur J Clin Investig 26:786–792

Blanco I, de Serres FJ, Fernandez-Bustillo E, Lara B, Miravitlles M (2006) Estimated numbers and prevalence of PI*S and PI*Z alleles of alpha1-antitrypsin deficiency in European countries. Eur Respir J 27 (21):77–84

Chorostowska-Wynimko J, Bakuła A, Kulus M, Kuca P, Niżankowska-Mogilnicka E, Sanak M, Socha P, Śliwiński P (2016) Standards for diagnosis and care of patients with inherited alpha-1 antitrypsin deficiency. Pneumonol Alergol Pol 84:193–202

Chorostowska-Wynimko J, Struniawski R, Sliwinski P, Wajda B, Czajkowska-Malinowska M (2015) The national alpha-1 antitrypsin deficiency registry in Poland. COPD 12(S1):22–26

Chorostowska-Wynimko J, Gawryluk D, Struniawski R, Popławska P, Fijołek J (2013) Incidence of alpha-1-antitrypsin Z and S alleles in patients with granulomatosis with polyangiitis – pilot study. Pneumonol Alergol Pol 81:319–322

Chorostowska-Wynimko J, Struniawski R, Popławska B, Borszewska-Kornacka M (2012) The incidence of alpha-1-antitrypsin (A1AT) deficiency alleles in population of Central Poland – preliminary results from newborn screening. Pneumonol Alergol Pol 80:450–453

Elzouki AN, Segelmark M, Wieslander J, Eriksson S (1994) Strong link between the alpha 1-antitrypsin PiZ allele and Wegener's granulomatosis. J Intern Med 236:543–548

Ferrarotti I, Thun GA, Zorzetto M, Ottaviani S, Imboden M, Schindler C, von Eckardstein A, Rohrer L, Rochat T, Russi EW, Probst-Hensch NM, Luisetti M (2012) Serum levels and genotype distribution of α1-antitrypsin in the general population. Thorax 67:669–674

Fregonese L, Stolk J (2008) Hereditary alpha-1-antitrypsin deficiency and its clinical consequences. Orphanet J Rare Dis 3:16

Gómez-Gómez A, Martínez-Martíne MU, Cuevas-Orta E, Bernal-Blanco JM, Cervantes-Ramírezd D, Martínez-Martínez R, Abud-Mendoza C (2014) Pulmonary manifestations of granulomatosis with polyangiitis. Rheumatic Clin 10:288–293

Hersh CP, Dahl M, Ly NP, Berkey CS, Nordestgaard BG, Silverman EK (2004) Chronic obstructive pulmonary disease in alpha1-antitrypsin PI MZ heterozygotes: a meta-analysis. Thorax 59:843–849

Janciauskiene SM, Bals R, Koczulla R, Vogelmeier C, Kohnlein T, Welte T (2011) The discovery of α1-antitrypsin and its role in health and disease. Respir Med 105:1129–1139

Kronbichler A, Kerschbaum J, Gründlinger G, Leierer J, Mayer G, Rudnicki M (2016) Evaluation and validation of biomarkers in granulomatosis with polyangiitis and microscopic polyangiitis. Nephrol Dial Transplant 31:930–936

Lara B, Martínez MT, Blanco I, Hernández-Moro C, Velasco EA, Ferrarotti I, Rodriguez-Frias F, Perez L, Vazquez I, Alonso J, Posada M, Martínez-Delgado B (2014) Severe alpha-1 antitrypsin deficiency in composite heterozygotes inheriting a new splicing mutation QOMadrid. Respir Res 15:125

Lyons PA, Rayner TF, Trivedi S et al (2012) Genetically distinct subsets within ANCA-associated vasculitis. N Engl J Med 367:214–223

Lutalo PM, D'Cruz DP (2014) Diagnosis and classification of granulomatosis with polyangiitis (aka Wegener's granulomatosis). J Autoimmun 48–49:94–98

Mahr AD, Edberg JC, Stone JH, Hoffman GS, St Clair WS, Specks U, Dellaripa PF, Seo P, Spiera RF, Rouhani FN, Brantly M, Merkel PA (2010) Alpha-1-antitrypsin deficiency-related alleles Z and S and the risk of Wegener's granulomatosis. Arthritis Rheum 62:3760–3767

McKinney EF, Willcocks LC, Broecker V, Smith KG (2014) The immunopathology of ANCA-associated vasculitis. Semin Immunopathol 36:461–478

Millet A, Pederzoli-Ribeil M, Guillevin L, Witko-Sarsat V, Mouthon L (2014) Antineutrophil cytoplasmic antibody-associated vasculitides: is it time to split up the group? Postgrad Med J 90:290–296

Morris H, Morgan MD, Wood AM, Smith SW, Ekeowa UI, Herrmann K, Holle JU, Guillevin L, Lomas DA, Perez J, Pusey CD, Salama AD, Stockley R, Wieczorek S, McKnight AJ, Maxwell AP, Miranda E, Williams J, Savage CO, Harper L (2011) ANCA-associated vasculitis is linked to carriage of the Z allele of α1-antitrypsin and its polymers. Ann Rheum Dis 70:1851–1856

Ottaviani SP, Gorrini M, Scabini R, Kadija Z, Paracchini E, Mariani F, Ferrarotti I, Luisetti M (2011) C reactive protein and alpha1-antitrypsin: relationship between levels and gene variants. Transl Res 157:332–338

Pervakova MY, Emanuel VL, Titova ON, Lapin SV, Mazurov VI, Belyaeva IB, Chudinov AL, Blinova TV, Surkova EA (2016) The diagnostic value of alpha-1-antitrypsin phenotype in patients with granulomatosis with polyangiitis. Int J Rheumatol 2016:7831410

Popławska B, Janciauskiene S, Chorostowska-Wynimko J (2013) Genetic variants of alpha-1 antitrypsin: classification and clinical implications. Pneumonol Alergol Pol 81(1):45–54

Stassen PM, Cohen-Tervaert JW, Lems SPM, Hepkema BG, Kallenberg CG, Stegeman CA (2009) HLA-DR4, DR13(6) and the ancestral haplotype A1B8DR3 are associated with ANCA-associated vasculitis and Wegener's granulomatosis. Rheumatology 48:622–625

Stockley RA, Turner AM (2014) α-1-Antitrypsin deficiency: clinical variability, assessment, and treatment. Trends Mol Med 20:105–115

Wiatr E, Gawryluk D (2013) Primary systemic ANCA-associated vasculitis-recommendations concerning diagnosis and treatment. Pneumonol Alergol Pol 81(5):479–491

Yates M, Watts RA, Bajema IM, Cid MC, Crestani B, Hauser T, Hellmich B, Holle JU, Laudien M, Little MA, Luqmani RA, Mahr AD, Merkel PA, Mills J, Mooney J, Segelmark M, Tesar V, Westman K, Vaglio A, Yalçındağ N, Jayne DR, Mukhtyar C (2016) EULAR/ERA-EDTA recommendations for the management of ANCA-associated vasculitis. Ann Rheum Dis 75(9):1583–1594. doi:10.1136/annrheumdis-2016-209133

Xiao H, Hu P, Falk RJ, Jennette JC (2016) Overview of the pathogenesis of ANCA-associated vasculitis. Kidney Dis (Basel) 1:205–215

Advs Exp. Medicine, Biology - Neuroscience and Respiration (2018) 39: 19–25
DOI 10.1007/5584_2018_152
© Springer International Publishing AG 2018
Published online: 20 Feb 2018

Hyperglycemia in Children Hospitalized with Acute Asthma

Khalid F. Mobaireek, Abdulrahman Alshehri,
Abdulaziz Alsadoun, Abdullah Alasmari, Abdullah Alashhab,
Meshal Alrumaih, Mohammad Alothman,
and Abdullah A. Alangari

Abstract

Hyperglycemia is frequently observed in adults with acute asthma. We aimed to assess the frequency of hyperglycemia and its relation to outcomes in children admitted with acute asthma. In this retrospective study, we reviewed medical records of non-diabetic 166 children (66 girls) with the mean age of 5.4 ± 2.6 years (range of 2–12 years), who were hospitalized with acute asthma between January 2012 through December 2014. Data pertaining to demographics, vital signs, oxygen saturation, serum blood glucose level, electrolytes, blood gases, and admission were collected. Children with other chronic conditions were excluded. The findings were that hyperglycemia (blood glucose ≥ 11.1 mmol/l) was observed in 38.6% of children. The median baseline blood glucose (IQR) was 9.8 mmol/l (7.2–13.3 mmol/l). Blood glucose level was associated with the length of hospitalization, with a median extension of 1.8 days, but was inversely associated with the serum potassium and bicarbonate levels. There were no associations between baseline blood glucose and age, gender, baseline respiratory rate, oxygen saturation, or intensive care admission. Hyperglycemia resolved spontaneously in all affected children. We conclude that hyperglycemia is common in children hospitalized with acute asthma. Hyperglycemia could be considered as a marker of a longer hospital stay.

Keywords

Acute asthma · Asthma exacerbation · Blood glucose · Children · Beta2-adrenergic agonists · Hyperglycemia

1 Introduction

Asthma is the most common chronic disease of childhood and a leading cause of emergency visits and hospitalization (Masoli et al. 2004) Hyperglycemia is known to develop in critically ill children with sepsis (Branco et al. 2005), viral illnesses such as bronchiolitis (Branco and Tasker 2007), and is also observed in patients with acute asthma (Koskela et al. 2013; Chawla et al. 2007).

K. F. Mobaireek, A. Alshehri, A. Alsadoun, A. Alasmari, A. Alashhab, M. Alrumaih, and A. A. Alangari (✉)
Department of Pediatrics, College of Medicine, King Saud University, Riyadh, Saudi Arabia
e-mail: aangari@ksu.edu.sa

M. Alothman
Department of Emergency Medicine, College of Medicine, King Saud University, Riyadh, Saudi Arabia

Several factors may contribute to the appearance of hyperglycemia, including stress induced by acute airway obstruction and the associated respiratory infections (Dungan et al. 2009). In addition, acute asthma therapy, particularly the use β_2-adrenergic agonists and systemic glucocorticoids, is conducive to the development of hyperglycemia (Dawson et al. 1995; Shires et al. 1979; Nogrady et al. 1977; Tickner et al. 1977). This transient hyperglycemia was once thought to be potentially beneficial to patients by providing glucose to organs and maintaining intravascular volume by increasing serum osmolality (Dungan et al. 2009; Schacke et al. 2002). However, subsequent reports have shown that hyperglycemia could be associated with worse clinical outcomes in adults with acute asthma, chronic obstructive pulmonary disease, sepsis, and other critical illnesses, as well as in children with acute viral bronchiolitis (Koskela et al. 2013; Hirshberg et al. 2008; Branco and Tasker 2007; Branco et al. 2005).

Despite the frequent observation of hyperglycemia in acute asthma, only few studies have so far addressed this complication (Koskela et al. 2013; Chawla et al. 2007; Majahan and Kamat 2007; Dawson et al. 1995; Smith et al. 1992). Data on the relation of hyperglycemia with the severity and outcomes in acute asthma are scarce in both children and adult populations. Therefore, in the present study we set out to retrospectively evaluate the frequency and outcomes of hyperglycemia in children admitted to the hospital with moderate-to-severe acute asthma.

2   Methods

2.1   Patients

The institutional review board of King Khalid University Hospital approved this study. A waiver of patients' consent was granted due to the retrospective nature of the study. We reviewed medical charts of all children between 2 and 12 years of age who were hospitalized with acute asthma exacerbation from January 2012 through December 2014 and who had a baseline serum glucose concentration measured within 4 h of arrival to the Emergency Department. Children with overt diabetes, chronic lung disease other than asthma, or other chronic cardiovascular, metabolic, renal, or neurological illnesses were excluded. Demographic and clinical data including age, gender, height, weight, basic clinical and laboratory information, medications given at home and the length of hospitalization were recorded in a standardized data collection form. Children with the baseline blood glucose level ≥ 11.1 mmol/l were identified as hyperglycemic according to the guidelines of the American Diabetes Association (2003) on the Diagnosis and Classification of Diabetes Mellitus and on the definition of hospital-related hyperglycemia.

2.2   Statistical Elaboration

Data were presented as means $\pm$SD or medians with 95% confidence intervals (95%CI). Fisher's exact test was used to compare the risk of hyperglycemia in boys and girls as well as the admissions to the Pediatric Ward and the Pediatric Intensive Care Unit. An unpaired t-test, with the Welch correction, and the Mann-Whitney U test were used to compare the length of hospitalization in normoglycemic and hyperglycemic patients. The Spearman correlation coefficient was used for the assessment of associations among the indices investigated. The evaluation was conducted with GraphPad Prism v6 software (San Diego, CA). A p-value < 0.05 defined statistically significant differences.

3   Results

A total of 313 children of 2–12 years of age were admitted to King Khalid University Hospital during the study period. There were 117 children with no measurement of the baseline blood glucose level or this level could not be retrieved from medical records at the time of data collection. In addition, 30 children were excluded because of other chronic systemic co-morbidities. Eventually, 166 children were included in the study.

Table 1 Characteristics of children with and without hyperglycemia (blood glucose $\geq$11.1 mmol/l)

	Hyperglycemics ($n = 64$)	Normoglycemics ($n = 102$)
Age (yr)	5.4 ± 2.7	5.4 ± 2.6
Body mass index (kg/m^2)	17.4 ± 5.6	17.8 ± 5.9
Respiratory rate (breaths/min)	41 ± 9.1	40 ± 15.3
O$_2$ saturation (%)	93.4 ± 3.7	92.6 ± 6.2
White blood cell count (x 10^9 per l)	13.7 ± 7.0	12.9 ± 6.3
Neutrophils (%)	79.0 ± 19.6	75.0 ± 16.9
Potassium (K$^+$) (mmol/l)	3.3 ± 0.5	3.6 ± 0.6
Bicarbonate (HCO$_3{}^-$) (mmol/l)	20.0 ± 4.3	21.9 ± 3.0

Data are means ±SD

There were 100 (60%) boys and 66 (40%) girls. The mean age of the entire cohort was 5.4 ± 2.6 years. The median baseline blood glucose was 9.8 mmol/l, interquartile range (IQR) 7.2–13.3 mmol/l, and min-max range of 4.2–20.3 mmol/l. The mean baseline blood glucose was 10.3 ± 3.8 mmol/l. Sixty-seven children had the measurement of blood glucose at Day 2, with a mean drop of 2.8 ± 4.7 mmol/l from Day 1, and 35 children had the measurement repeated at Day 3, with a mean drop of 1.9 ± 4.5 mmol/l from Day 2. Table 1 shows the characteristics of the children with and without hyperglycemia. Sixty-four (38.6%) children had baseline blood glucose level $\geq$ 11.1 mmol/l, i.e., above the normal limit according to the guidelines of the American Diabetes Association (2003). There was no significant difference between boys and girls in the rate of development of hyperglycemia, OR = 1.17 (95%CI: 0.61–2.22; p = 0.75) (Fig. 1), nor was there any association between the baseline blood glucose level and age, body mass index (BMI), baseline respiratory rate, baseline oxygen saturation, white blood cell count, or the neutrophil count. Forty-six children were admitted to the Pediatric Intensive Care Unit and 120 admitted to the Pediatric Ward. Hyperglycemia was observed in 33% of the former admissions and in 41% of the latter admissions, OR = 0.7 (95% CI: 0.34–1.44) (Fig. 2).

The baseline blood glucose level was associated with the length of hospitalization (Table 2). The mean hospitalization time amounted to 4.9 ± 3.0 days for all children. It increased to 6.0 ± 3.5 days for children with baseline hyperglycemia, while it amounted to 4.2 ± 2.4 days for those with normoglycemia. The difference between the mean hospitalization time in the two groups of children amounted to 1.8 days (95%CI: 0.8–2.9) and was significant (p = 0.0003). However, the length of hospitalization did not associate with the blood glucose level at Day 2 (r^2 = 0.08; 95% CI: -0.18–0.32) (p = 0.55) or Day 3 (r^2 = −0.17; 95%CI: -0.49–0.19) (p = 0.34). We found, however, that the baseline blood glucose level was inversely associated with both serum K$^+$ and HCO$_3{}^-$ levels (Table 2). Both K$^+$ and HCO$_3{}^-$ positively associated with each other, but neither associated with the length of hospitalization.

4 Discussion

This study demonstrates that hyperglycemia was common in children treated in the Emergency Department and subsequently admitted with moderate-to-severe acute asthma. Hyperglycemia has been reported in 79% of adults with acute asthma receiving nebulized β_2-agonists and steroids (Koskela et al. 2013). β_2-agonists may cause hyperglycemia by enhancing hepatic and muscle glycogenolysis and gluconeogenesis (Chawla et al. 2007). These drugs also release fatty acids by enhancing lipolysis, which causes acute insulin resistance (Dresner et al. 1999). Moreover, the use of β_2-agonists is associated with hypokalemia and metabolic acidosis due to increased lactate production (Meert et al. 2012;

Fig. 1 Number of girls and
boys with normoglycemia
and hyperglycemia

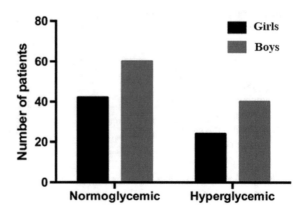

Fig. 2 Number of children
hospitalized with
normoglycemia and
hyperglycemia in the
Pediatric Ward and the
Pediatric Intensive
Care Unit

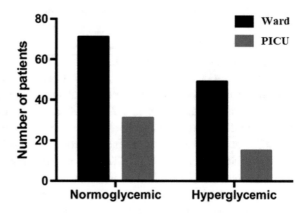

Table 2 Associations between baseline blood glucose (BG) and hospitalization length (HL), serum potassium (K^+), and serum bicarbonate (HCO_3^-)

	BG vs. HL	BG vs. K^+	BG vs. HCO_3^-	K^+ vs. HCO_3^-
Correlation coefficient (r^2)	0.36	−0.37	−0.31	0.2
95% CI	0.21 to 0.49	−0.5 to −0.22	−0.45 to −0.15	0.03 to 0.35
p-value	< 0.0001	< 0.0001	0.0003	0.0180

Meert et al. 2007; Alberts et al. 1986). The present findings show an inverse association between K^+ and HCO_3^- levels and baseline blood glucose, which is in line with the notion that hyperglycemia may be secondary to the use of β_2-agonists. This notion is further supported by a gradual decrement in the mean blood glucose level at Day 2 and Day 3, which usually coincided with less frequent β_2-agonist dosing.

The contribution of steroids to hyperglycemia is less distinct than that of β_2-agonists and takes a longer time to develop (Smith et al. 1992). In the present population of children, blood glucose level was measured within only few hours after their arrival to the Emergency Department and it would be unlikely for one dose of systemic steroids to cause that rapid a surge in blood glucose. Stress could also induce hyperglycemia by increasing cortisol, glucagon, growth hormone, catecholamines, and various cytokines, all of which may stimulate gluconeogenesis and glycogenolysis (Dungan et al. 2009). We failed to find any association between hyperglycemia and acute asthma severity as assessed by baseline respiratory rate, oxygen saturation, and admission to the Pediatric Intensive Care Unit. Further, hyperglycemia is rare in community-acquired pneumonia regardless of its severity (Don et al. 2008), suggesting that stress might not be a major contributing factor to hyperglycemia in acute asthma. On the other hand, the fasting state of patients who present with acute asthma delays the insulin-mediated glucose uptake by the liver, which exaggerates and prolongs the hyperglycemia (DeFronzo et al. 1978).

In the present study we found a significant association between hyperglycemia and increased duration of hospitalization. The presence of hyperglycemia was associated with almost 2-day-longer hospitalization. This, as far as we know, has never been reported in children. Wytrychowski et al. (2016) have found that hyperglycemia is a significant factor in extending hospitalization time in adults with acute asthma. However, those authors investigated a small mixed population sample consisting of diabetics and non-diabetics. On the other hand, hyperglycemia has been associated with longer intensive care stay and ventilation in children with viral bronchiolitis (Branco and Tasker 2007). That suggests a possible relationship between hyperglycemia and persistense of airway inflammation or a decrease in effectiveness of asthma therapy. Hyperglycemia has been shown *in vitro* to upregulate several cytokines, such as tumor necrosis factor-α, interleukin-1, and interleukin-6 (Ling et al. 2005; Esposito et al. 2002). These cytokines could further exacerbate the inflammatory process in the airways. Further, acidosis, mostly induced by β_2-agonists, reduces the bronchodilatory effect of β_2-agonists (Tenney 1956). It is rather inexplicable why only 38.6% children of the present study developed hyperglycemia, although they had been treated in a similar way as the other children. Individual genetic variations that predispose to hyperglycemia might play a role (Dungan et al. 2009; Schacke et al. 2002). The β_2-adrenergic receptor gene has multiple polymorphisms that can alter the response to drugs (Taylor 2007; Evans and Johnson 2001).

Hyperglycemia in critically ill children is generally associated with increased mortality and morbidity (Hirshberg et al. 2008). However, there is no evidence thus far that control of blood glucose by insulin will improve the outcome in this population (Agus et al. 2012; Vlasselaers et al. 2009). A meta-analysis of critically ill adults (Griesdale et al. 2009) and two randomized cohort studies (Macrae et al. 2014; Van den Berghe et al. 2006) have failed to show a benefit of insulin therapy, and a large multicenter trial has shown that tight glycemic control, unexpectedly, increases mortality (Wiener et al. 2008). Cochius-den Otter et al. (2015) have observed in children admitted to the Pediatric Intensive Care Unit with severe acute asthma and hyperglycemia that therapeutic use of intravenous insulin to control hyperglycemia does not reduce the duration of intravenous salbutamol infusion or hospitalization time. Likewise, Wytrychowski et al. (2016) have found that increased hospital time in adults with acute asthma and hyperglycemia is not affected by insulin therapy.

The main limitation of this study was its retrospective nature. The timing of baseline blood glucose measurements could not be strictly standardized, so that the relationship of blood glucose level to treatments given in the Emergency Department could not be accurately assessed, although the samples for baseline blood glucose were mostly taken within 4 h of presentation. Data concerning the treatment used at home, including β_2-agonists or steroids, were also lacking.

In conclusion, hyperglycemia is common in children with acute asthma and it is likely to be due to the therapeutic use of β_2-agonists. Although hyperglycemia is not related to asthma severity, it is associated with increased hospitalization time. There is no evidence from the current literature to support the use of insulin to treat hyperglycemia to improve acute asthma outcomes. Prospective studies specially designed to decipher the mechanisms of a link between hyperglycemia and increased hospitalization time may help improve the management of children with acute asthma and may save costs.

Acknowledgements The authors extend their appreciation to the Deanship of Scientific Research at King Saud University for funding this work through the research grant No. RGP-190.

Conflicts of Interest The authors declare that they have no conflict of interest in relation to this article.

References

Agus MS, Steil GM, Wypij D, Costello JM, Laussen PC, Langer M, Alexander JL, Scoppettuolo LA, Pigula FA, Charpie JR, Ohye RG, Gaies MG, SPECS Study Investigators (2012) Tight glycemic control versus standard care after pediatric cardiac surgery. N Engl J Med 367(13):1208–1219

Alberts WM, Williams JH, Ramsdell JW (1986) Metabolic acidosis as a presenting feature in acute asthma. Ann Allergy 57(2):107–109

American Diabetes Association (2003) Expert committee on the diagnosis and classification of diabetes mellitus. Diabetes Care 26(Suppl 1):S5–20

Branco RG, Tasker RC (2007) Glycemic level in mechanically ventilated children with bronchiolitis. Pediatr Crit Care Med 8(6):546–550

Branco RG, Garcia PC, Piva JP, Casartelli CH, Seibel V, Tasker RC (2005) Glucose level and risk of mortality in pediatric septic shock. Pediatr Crit Care Med 6 (4):470–472

Chawla S, Seth D, Cortez J (2007) Asthma and hyperglycemia. Patient report. Clin Pediatr 46(5):454–455

Cochius-den Otter SC, Joosten KF, de Jongste JC, Hop WC, de Hoog M, Buysse CM (2015) Insulin therapy in hyperglycemic children with severe acute asthma. J Asthma 52(7):681–686

Dawson KP, Penna AC, Manglick P (1995) Acute asthma, salbutamol and hyperglycaemia. Acta Paediatr 84 (3):305–307

DeFronzo RA, Ferrannini E, Hendler R, Wahren J, Felig P (1978) Influence of hyperinsulinemia, hyperglycemia, and the route of glucose administration on splanchnic glucose exchange. Proc Natl Acad Sci U S A 75 (10):5173–5177

Don M, Valerio G, Korppi M, Canciani M (2008) Hyper- and hypoglycemia in children with community-acquired pneumonia. J Pediatr Endocrinol Metab 21 (7):657–664

Dresner A, Laurent D, Marcucci M, Griffin ME, Dufour S, Cline GW, Slezak LA, Andersen DK, Hundal RS, Rothman DL, Petersen KF, Shulman GI (1999) Effects of free fatty acids on glucose transport and IRS-1-associated phosphatidylinositol 3-kinase activity. J Clin Invest 103(2):253–259

Dungan KM, Braithwaite SS, Preiser JC (2009) Stress hyperglycaemia. Lancet 373(9677):1798–1807

Esposito K, Nappo F, Marfella R, Giugliano G, Giugliano F, Ciotola M, Quagliaro L, Ceriello A, Giugliano D (2002) Inflammatory cytokine concentrations are acutely increased by hyperglycemia in humans: role of oxidative stress. Circulation 106(16):2067–2072

Evans WE, Johnson JA (2001) Pharmacogenomics: the inherited basis for interindividual differences in drug response. Annu Rev Genomics Hum Genet 2:9–39

Griesdale DE, de Souza RJ, van Dam RM, Heyland DK, Cook DJ, Malhotra A, Dhaliwal R, Henderson WR, Chittock DR, Finfer S, Talmor D (2009) Intensive insulin therapy and mortality among critically ill patients: a meta-analysis including NICE-SUGAR study data. CMAJ 180(8):821–827

Hirshberg E, Larsen G, Van Duker H (2008) Alterations in glucose homeostasis in the pediatric intensive care unit: hyperglycemia and glucose variability are associated with increased mortality and morbidity. Pediatr Crit Care Med 9(4):361–336

Koskela HO, Salonen PH, Niskanen L (2013) Hyperglycaemia during exacerbations of asthma and chronic obstructive pulmonary disease. Clin Respir J 7 (4):382–389

Ling PR, Smith RJ, Bistrian BR (2005) Hyperglycemia enhances the cytokine production and oxidative responses to a low but not high dose of endotoxin in rats. Crit Care Med 33(5):1084–1089

Macrae D, Grieve R, Allen E, Sadique Z, Morris K, Pappachan J, Parslow R, Tasker RC, Elbourne D, CHiP Investigators (2014) A randomized trial of hyperglycemic control in pediatric intensive care. N Engl J Med 370(2):107–118

Majahan P, Kamat D (2007) Asthma and hyperglycemia. Diagnosis: acute asthma exacerbation worsened by hyperglycemia. Clin Pediatr (Phila) 46(5):455–457

Masoli M, Fabian D, Holt S, Beasley R (2004) The global burden of asthma: executive summary of the GINA dissemination committee report. Allergy 59 (5):469–478

Meert KL, Clark J, Sarnaik AP (2007) Metabolic acidosis as an underlying mechanism of respiratory distress in children with severe acute asthma. Pediatr Crit Care Med 8(6):519–523

Meert KL, McCaulley L, Sarnaik AP (2012) Mechanism of lactic acidosis in children with acute severe asthma. Pediatr Crit Care Med 13(1):28–31

Nogrady SG, Hartley JP, Seaton A (1977) Metabolic effects of intravenous salbutamol in the course of acute severe asthma. Thorax 32(5):559–562

Schacke H, Docke WD, Asadullah K (2002) Mechanisms involved in the side effects of glucocorticoids. Pharmacol Ther 96(1):23–43

Shires R, Joffe BI, Heding LG, Seftel HC (1979) Metabolic studies in acute asthma before and after treatment. Br J Dis Chest 73(1):66–70

Smith AP, Banks J, Buchanan K, Cheong B, Gunawardena KA (1992) Mechanisms of abnormal glucose metabolism during the treatment of acute severe asthma. Q J Med 82(297):71–80

Taylor MR (2007) Pharmacogenetics of the human beta-adrenergic receptors. Pharmacogenomics J 7(1):29–37

Tenney SM (1956) Sympatho-adrenal stimulation by carbon dioxide and the inhibitory effect of carbonic acid on epinephrine response. Am J Phys 187(2):341–346

Tickner TR, Cramp DG, Foo AY, Johnson AJ, Bateman SM, Pidgeon J, Spiro SG, Clarke SW, Wills MR (1977) Metabolic response to intravenous salbutamol therapy in acute asthma. Thorax 32(2):182–184

Van den Berghe G, Wilmer A, Hermans G, Meersseman W, Wouters PJ, Milants I, Van Wijngaerden E, Bobbaers H, Bouillon R (2006) Intensive insulin therapy in the medical ICU. N Engl J Med 354(5):449–461

Vlasselaers D, Milants I, Desmet L, Wouters PJ, Vanhorebeek I, van den Heuvel I, Mesotten D, Casaer MP, Meyfroidt G, Ingels C, Muller J, Van Cromphaut S, Schetz M, Van den Berghe G (2009) Intensive insulin therapy for patients in paediatric intensive care: a prospective, randomised controlled study. Lancet 373(9663):547–556

Wiener RS, Wiener DC, Larson RJ (2008) Benefits and risks of tight glucose control in critically ill adults: a meta-analysis. JAMA 300(8):933–944

Wytrychowski K, Obojski A, Hans-Wytrychowska A (2016) The influence of insulin therapy on the course of acute exacerbation of bronchial asthma. Adv Exp Med Biol 884:45–51

Advs Exp. Medicine, Biology - Neuroscience and Respiration (2018) 39: 27–36
https://doi.org/10.1007/5584_2018_157
© Springer International Publishing AG 2018
Published online: 22 March 2018

Serum Vitamin D Concentration and Markers of Bone Metabolism in Perimenopausal and Postmenopausal Women with Asthma and COPD

K. Białek-Gosk, R. Rubinsztajn, S. Białek, M. Paplińska-Goryca, R. Krenke, and R. Chazan

Abstract

Aging and menopause are closely related to hormonal and metabolic changes. Vitamin D is a crucial factor modulating several metabolic processes. The aim of this study was to evaluate biomarkers of bone metabolism in peri- and postmenopausal women with obstructive lung diseases. Sixty two female patients, 27 with asthma and 35 with COPD, aged over 45 years (median age 58 and 64 years, respectively) were enrolled into the study. The evaluation included lung function, bone mineral density, serum concentration of vitamin D, and bone metabolism markers. The study groups differed significantly in terms of forced expiratory volume in 1 s (FEV_1); median values of 1.79 L $vs.$ 1.16 L (p = 0.0001) and 71.2% vs. 53.0% predicted (p = 0.0072) and in vitamin D concentration (12.3 ng/ml vs. 17.6 ng/ml). Total bone mineral density (BMD) was lower in the COPD group (p = 0.0115). Serum vitamin D inversely correlated with the number of pack-years in asthma patients (r = −0.45, p = 0.0192). There was no correlation between serum vitamin D and disease duration or severity, and the Asthma Control Test (ACT) and the modified Medical Research Council (mMRC) dyspnea scores. The serum bone metabolism markers C-terminal cross-linked telopeptide of collagen type I (BCROSS), N-terminal propeptides of procollagen type-1 (tP1NP), and N-mid osteocalcin (OCN) inversely correlated with age in the COPD, but not asthma, patients (r = −0.38, p = 0.0264; r = −0.37, p = 0.0270; and r = −0.42, p = 0.0125, respectively). We conclude that peri- and postmenopausal women with obstructive lung diseases had a decreased serum concentration of vitamin D. Furthermore, vitamin D and body mineral density were appreciably lower in women with COPD than those with asthma.

Keywords

Asthma · Bone metabolism markers · Bone mineral density · COPD · Menopause · Osteoporosis · Pulmonary function · Vitamin D

K. Białek-Gosk, R. Rubinsztajn (✉),
M. Paplińska-Goryca, R. Krenke, and R. Chazan
Department of Internal Medicine, Pulmonary Diseases and Allergy, Medical University of Warsaw, Warsaw, Poland
e-mail: rrubinsztajn@wum.edu.pl

S. Białek
Department of Biochemistry and Clinical Chemistry, Medical University of Warsaw, Warsaw, Poland

1   Introduction

Aging entails a number of changes in the female body. These changes are closely related to a reduced estrogen production by ovaries and a significant decrease in the serum estrogen level and calcium absorption. Estrogen inhibits bone resorption and stimulates bone formation. Thus, estrogen and calcium deficiency results in a reduction of bone mass. 25-hydroxylated vitamin D (25(OH)D) is yet another critical factor modulating bone growth and turnover. Under normal condition, vitamin D is responsible for maintaining a proper calcium level, mainly by promoting the intestinal calcium absorption. Aging is associated with a decrease in skin synthesis of vitamin D3 and impaired production of its active metabolite in the kidney (Veldurthy et al. 2016). That may result in decreased calcium absorption from the gastrointestinal tract promoting osteoporosis and increasing susceptibility to bone fractures. Recently, there has been a growing interest in the role of vitamin D in the development and treatment of respiratory diseases, including asthma. Some promising reports on the possible therapeutic applications of vitamin D in asthma patients have been published (Jiao et al. 2016). The beneficial effects of vitamin D can be explained by several mechanisms. Firstly, vitamin D decreases the production of inflammatory cytokines by T-helper type-9 lymphocytes, such as interleukin-5 (IL-5), IL-9, and IL-13, which play an important role in the pathogenesis of asthma (Keating et al. 2014). Secondly, steroid resistance, a major problem in asthma treatment, can be overcome by vitamin D which activates IL-10 secreting regulatory T cells and down-regulates the expression of fractalkine, a chemokine underlying steroid resistance. Interestingly, there is an overlap between the risk factors for vitamin D deficiency and asthma development. These include: residing in urban areas, obesity and ethnicity. Some links between vitamin D, bone mineral density and chronic obstructive pulmonary disease (COPD) have also been noted. Although the pathogenesis of COPD is mainly related to inflammation, oxidative stress, pulmonary protease-antiprotease imbalance, impaired lung development in early life has been identified as a potential risk factor for COPD. The results of experimental studies suggest the role of vitamin D in the growth and development. The presence of vitamin D in alveolar type II cells increases surfactant synthesis and regulates epithelial-mesenchymal interactions. A relationship between smoking and vitamin D signaling has also been demonstrated. Cigarette smoke extracts inhibit vitamin D receptor translocation in human alveolar epithelial cells and causes down-regulation of local vitamin D signaling, which leads to the insufficient control of pro-inflammatory processes in the airways of COPD patients (Janssens et al. 2013). Smoking is associated with a significant reduction in bone mineral density and it is a risk factor for osteoporosis. It should be underlined that in both asthma and COPD steroid treatment may affect bone metabolism and lead to osteoporotic fractures. Considering the complex relationships between aging, menopause, vitamin D, bone mineral density, and asthma and COPD, we undertook a study to evaluate the parameters characterizing bone metabolism in perimenopausal and postmenopausal women with obstructive lung diseases.

2   Methods

2.1   Patients and Study Design

This was a prospective, cross-sectional study performed at the Medical University of Warsaw, Poland, between May 2011 and June 2012 in peri- or postmenopausal women with asthma or COPD. The study was approved by the Institutional Review Board for Human Research. Patients were recruited from the out-patient clinic of the Department of Internal Medicine, Pulmonary Diseases and Allergy of Warsaw Medical University. All patients gave written informed consent to participate in the study. Sixty-two women (37 with asthma and 35 with COPD) were included. The main inclusion criteria were as follows: at least 45 years of age, according to the assumptive staging of reproductive age (Harlow et al. 2012; Kaczmarek 2007) and asthma or COPD diagnosed at least 1 year before enrolment in the study. The exclusion criteria were as follows: exacerbations over the preceding 3 months, bisphosphonate or vitamin D therapy over the preceding 6 months, oral steroids over the preceding 6 weeks, use of hormonal replacement therapy and current osteoporosis treatment with bisphosphonates. All women lived at the same geographical latitude and thus had a similar exposure to sunlight.

Basic laboratory data, which were collected during the study, consisted of bone mineral density (BMD), serum vitamin D, C-terminal cross-linked telopeptide of collagen type I (BCROSS), N-terminal propeptides of procollagen type-1 (P1NP), and N-mid osteocalcin (OCN). Asthma and COPD were assessed from the medical history, physical examination, lung function, body mass index (BMI), and laboratory data. Smoking exposure was quantified in pack-years. Spirometry, with bronchial reversibility testing after administration of 400 µg salbutamol *via* a spacer (Lung test 1000; MES, Cracow, Poland), was performed in accordance with the recommendations of the European Respiratory Society (ERS) (Pellegrino et al. 2005).

2.2 Assessment of Disease Control

The Polish version of the Asthma Control Test (ACT) was used with the permission of the copyright owner to assess the disease symptoms, according to Global Initiative for Asthma recommendations (GINA 2010). The ACT encompasses the severity and frequency of asthma symptoms during the day and night, the use of reliever medications, and the disease-related limitation of daily activity. The score ranges from 5 to 25 points, with the higher score indicating better disease control. In addition, the perception of dyspnea was evaluated.

Asthma severity was assessed from a retrospective review of treatment. Following the GINA recommendations, mild asthma was considered when the disease control required reliever drugs only, leukotriene antagonists, or

low-dose inhaled steroids. Moderate asthma was diagnosed when the disease was controlled by low doses of inhaled steroids and long-acting beta-agonists (LABA), and severe asthma when high doses of inhaled steroids and LABA, or additional treatments were required.

COPD severity was assessed by the degree of airflow limitation in accordance with the Global Initiative for Obstructive Lung Disease criteria. The modified Medical Research Council (mMRC) scale was used for the assessment of dyspnea.

2.3 Laboratory Investigations

Fasting blood samples were taken from the antecubital vein between 7:30 and 9:30 a.m. The samples were centrifuged and serum was kept frozen at -75°C until assayed. Vitamin D, BCROSS, P1NP, and OCN were measured with high-sensitivity electrochemiluminescence immunoassays in an Elecsys 2010 automatic analyzer (Roche Diagnostics GmbH; Mannheim, Germany).

BMD was assessed at the postero-anterior lumbar spine level of L1-L4 and the femur neck bone using dual-energy X-ray absorptiometry (DEXA) (Discovery Densitometer; Hologic; Waltham, MA), according to the manufacturer's instruction. The baseline BMD was assigned the value of 100% and was compared with the final BMD result, expressed as a percentage difference from baseline. A Z-score was calculated for individual patients, expressed as the multiplication of standard deviations from an ideal BMD value corrected for age and sex.

2.4 Statistical Analysis

The results were expressed as the median values and inter-quartile range (IQR). Most of the variables failed to meet the requirement of normal distribution; therefore the Mann-Whitney U and Kruskal-Wallis tests were used for comparisons between continuous variables in unrelated groups. The Chi-squared test was used to assess the proportion of patients with different levels of

T-scores in the asthma and COPD groups. Correlations between parameters were checked with Spearman's correlation coefficient. A p value <0.05 defined statistically significant differences. All analyses were performed using a commercial statistical package of STATISTICA v8.0 (StatSoft; Tulsa, OK, USA).

3   Results

The characteristics of the study groups are presented in Table 1 and the spirometry results are summarized in Table 2.

Total bone mineral density was lower in COPD than in asthma patients; T-score −2,5 was found in 19 women with asthma and 14 with COPD (p = 0.0488). Except lower vitamin D levels in COPD patients, no other significant differences in the serum concentration of bone metabolism markers were found between asthma and COPD patients (Table 3).

In asthma, serum vitamin D concentration was inversely associated with the number of pack-years ($r = -0.45$, p = 0.0192) (Fig. 1). Also, total bone mineral density was inversely associated with the patient age ($r = -0.42$, p < 0.031) (Fig. 2). There was no association between the serum vitamin D concentration and disease duration, severity, and the ACT or mMRC score. In COPD, levels of the bone metabolism markers BCROSS, tP1NP, and OCN were associated with age (r = −0.38, p = 0.0264; $r = -0.37$, p = 0.0270; and r = −0.42, p = 0.0125, respectively). In asthma, total bone mineral density was inversely correlated with age (Fig. 2).

In COPD, but not in asthma, there was an inverse correlation between total BMD and mMRC (r = −0.37, p = 0.0389) (Fig. 3). We failed to find any relationship between disease duration (asthma or COPD) and the following variables: total BMD and vitamin D, BCROSS, tP1NP, and OCN. A positive correlation between BMI and total BMD was demonstrated in both asthma and COPD

Table 1 Characteristics of the study groups

	Asthma	COPD	
	(n = 27)	(n = 35)	p
Age (years)	58.0 (56.5–60.0)	64.0 (59.5–72.5)	0.0011
Disease duration (years)	15.0 (4.5–30.0)	2.0 (1.0–3.0)	<0.0001
BMI (kg/m²)	27.5 (25.6–32.0)	27.0 (23.0–29.5)	ns
Smoking status:			
Never smoker	13	1	0.0001
Current smoker	4	4	ns
Ex-smoker	10	30	0.0002
Cigarettes (pack-years)	8	35	<0.0001
ACT (score)	15.9	–	–
mMRC (score)	–	2.0	–
Patients treated with ICS (%)	27 (100)	20 (57)	0.0001

Data are medians (IQR). *ACT* Asthma Control Test, *mMRC* modified Medical Research Council scale for dyspnea, *ICS* inhaled corticosteroids

Table 2 Differences in spirometry results between the asthma and COPD patient groups

	Asthma	COPD	
	(n = 27)	(n = 35)	p
FEV$_1$ (L)	1.79 (1.56–2.01)	1.16 (0.84–1.60)	0.0001
FEV$_1$ (%pred)	71.2 (62.4–84.1)	53.0 (41.2–76.4)	0.0072
FVC (L)	2.87 (2.57–3.16)	2.50 (2.00–2.80)	0.0270
FVC (%pred)	94.9 (80.0–102.2)	90.7 (81.0–102.1)	ns
FEV$_1$%FVC	80.0 (70.0–84.1)	48.0 (42.0–59.5)	0.0281

Data are as medians (IQR), *FEV$_1$* forced expiratory volume in one second, *FVC* forced vital capacity

Table 3 Differences in bone mineral density and bone metabolism biomarkers between asthma and COPD patients

| | Asthma | COPD | |
	($n = 27$)	($n = 35$)	p
Total BMD (g/cm^2)	0.85 (0.77–0.97)	0.78 (0.68–0.89)	0.0115
T-score	−0.8 (−1.4–0.2)	−1.2 (−2.2–0.3)	0.0488
Z-score	0.2 (−0.5–0.9)	0.3 (−0.5–0.8)	ns
Femur neck BMD (g/cm^2)	0.70 (0.66–0.77)	0.68 (0.58–0.76)	ns
T-score	−1.3 (−1.7–0.6)	−1.4 (−2.4–0.7)	ns
Z-score	0.1 (−0.1–0.6)	0.1 (−0.4–0.6)	ns
L1-L2 BMD (g/cm^2)	0.85 (0.82–0.10)	0.88 (0.77–0.97)	ns
T-score	−1.8 (−2.1–0.4)	−1.5 (−2.5–0.6)	ns
Z-score	−0.2 (−0.75–0.7)	0.4 (−0.9–1.0)	ns
25(OH)D (ng/ml)	17.6 (14.2–20.1)	12.3 (8.9–15.6)	0.0011
BCROSS (pg/ml)	405 (301–615)	369 (201–513)	ns
tP1NP (ng/ml)	58.1 (37.8–73.8)	43.3 (29.3–58.9)	ns
Osteocalcin (ng/ml)	22.0 (16.0–32.3)	20.9 (12.9–27.4)	ns

Results are medians (IQR). *BMD* bone mineral density, *BCROSS* C-terminal cross-linked telopeptide of collagen type-1, *tP1NP* N-terminal propeptides of procollagen type-1, *25(OH)D* 25-hydroxylated vitamin D

Fig. 1 Correlation between concentrations of 25(OH)D and the number of pack-years in asthmatics

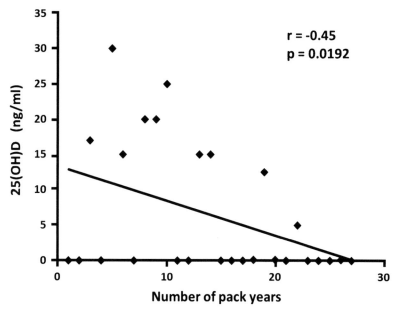

patients (Fig. 4). There was no association between the bone metabolism markers and lung function tests either in asthma of COPD.

4   Discussion

This study demonstrates that serum vitamin D concentration is decreased in both asthma and COPD female patients compared to the

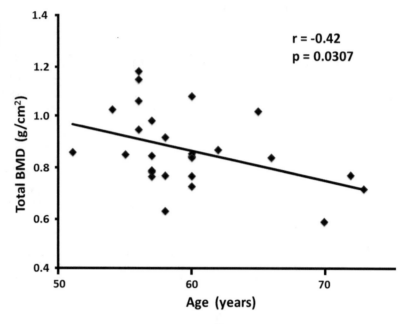

Fig. 2 Correlation between total bone mineral density (BMD) and patient age in asthmatics

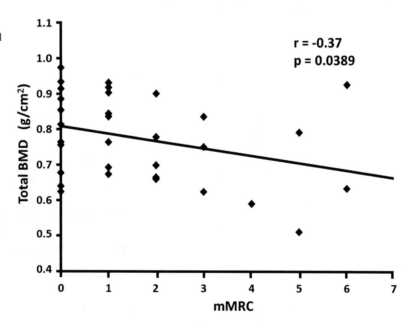

Fig. 3 Correlation between total bone mineral density (BMD) and modified medical research council (mMRC) scale in patients with chronic obstructive pulmonary disease (COPD)

reference values in the healthy population in Poland (Płudowski et al. 2013). In addition, vitamin D level was significantly lower in COPD than that in asthma patients. Likewise, total BMD was lower in COPD than that in asthma patients. A negative correlation between cigarette smoke exposure and serum vitamin D concentration was present in asthmatics but not in COPD patients. There were insignificant differences in the bone

Fig. 4 Correlation between body mass index (BMI) and total bone mineral density (BMD) in the asthma (panel A) and COPD (panel B) groups of patients

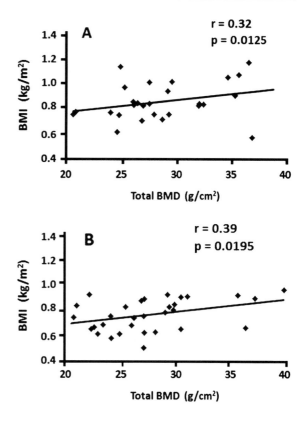

turnover markers in both groups of patients. Nor was there any relation between the results of lung function tests and serum vitamin D or bone turnover markers. The finding of lower vitamin D concentration in patients with obstructive lung diseases is in line with previous studies. Persson et al. (2012) have reported that COPD is associated with an increased risk of vitamin D deficiency and this is particularly relevant in current smokers, obese patients, and those with severe and very severe disease. Moberg et al. (2014) have reported a relationship between age and vitamin D deficiency in COPD patients. We failed to confirm such a correlation in the present study, but both age and the number of COPD patients were greater in the former study. Odler et al. (2015) have found that vitamin D concentration in the asthma-COPD overlap syndrome and in COPD are significantly lower than that in asthmatics, what suggests that vitamin D deficiency may affect control of this disease.

Confino-Cohen et al. (2014) have confirmed that low vitamin D concentration is associated with asthma exacerbation in the Israeli adult population. There is evidence that severity of airway obstruction and the course of asthma are associated with vitamin D deficiency in both children (Esfandiar et al. 2016) and adults (Korn et al. 2013). The association of a decline in vitamin D concentration with changes in various clinical indicators has been reported in both asthma and COPD. Low vitamin D content is linked to impaired lung function, greater frequency of exacerbations, and a reduced response to inhaled steroids (Cassim et al. 2015). In this study we failed to find a relation between low serum vitamin D concentration and airway obstruction in adult female asthma patients, what is in line with the observation of Larose et al. (2015). Brumpton et al. (2016) have shown that low vitamin D concentration is only weakly associated with a significant decline in lung function in adult asthmatics.

This association was stronger in never-smokers and in patients not treated with ICS. Another study not only demonstrates the negative impact of decreased vitamin D concentration on lung function and airway hyperresponsiveness in asthma, but also suggests a link between low vitamin D and reduced response to inhaled steroids. A relationship between serum vitamin D and FEV_1 and FVC has been documented and it has been shown that vitamin D supplementation as add-on therapy may improve FEV_1 (Arshi et al. 2014). However, vitamin D supplementation has not been associated with improvement in asthma control or reduction in exacerbation frequency. Similalry to asthma, some studies in COPD patients demonstrate that low serum vitamin D concentration is associated with a lower FEV_1 (Zendedel et al. 2015), disease severity (Janssens et al. 2011), and more severe exacerbations (Martineau et al. 2015). A relationship between low serum vitamin D concentration and obstructive pattern of lung function has been observed not only in asthma or COPD, but also in the general population. The odds ratio for the development of FEV1/FVC <0.7 in ever-smokers is greater in subjects with a low vitamin D level (Larose et al. 2015).

The present study demonstrates that similarly to vitamin D, serum total BMD was also significantly lower in COPD compared to asthma patients. It is known that the prevalence of osteoporosis is associated with age and smoking in perimenopausal women. In our patients with asthma and COPD, the effect of inhaled steroids on BMD should also be considered. There is a well-recognized relationship between treatment with inhaled steroids in asthma and COPD and the risk of osteoporosis and its complications, such as hip or vertebral fracture (Sweeney et al. 2016). Corticosteroids reduce bone formation and increase bone resorption by a direct effect on osteoblasts and osteoclasts. Female users of corticosteroids have a lower BMD at the mid-shaft radius, hip, and spine than never-users. Interestingly, such an association between BMD and corticosteroid use was not found in men. As the use of corticosteroids in asthma is closely related to disease severity, it might be difficult to evaluate the pure effect of asthma severity on the prevalence of osteoporosis. In a British study, the prevalence of osteoporosis in moderate/mild and severe asthma was 4% and 16%, respectively (Sweeney et al. 2016). Li et al. (2015) have reviewed the frequency of osteopenia/osteoporosis in moderate-to-severe asthma in patients treated with different doses of continuous or intermittent systemic and inhaled steroids. The authors found no difference in the prevalence of osteopenia/osteoporosis between continuous an intermittent systemic steroid treatment, but a difference between continuous systemic steroids and inhaled steroids treatment, to the disadvantage of the former. The authors found no significant differences in BMD between asthma patients treated with inhaled or systemic steroids compared to control subjects. Nevertheless, the prevalence of osteoporosis tended to be higher in the users of systemic steroids. COPD is currently considered a systemic disease, and according to the GOLD guidelines, osteoporosis is one of the important comorbidities (GOLD 2010). The National Health and Nutrition Examination Survey (NHANES) has reported a 17% prevalence of osteoporosis in COPD patients compared with 8.5% in subjects without COPD (Schnell et al. 2012). In a study of Silva et al. (2011) as many as 42% of patients were osteoporotic. There was no significant difference between the percentage of osteopenia/osteoporosis and normal-bone-mass patients in different GOLD stages, but the BODE index was higher in the osteoporosis patients. In a study of Graat-Verboom et al. (2012), the prevalence of osteoporosis during a three-year follow-up increased from 47% to 61%. In that study, trochanter osteopenia and vitamin D deficiency appeared to be the baseline risk factors for the development of osteoporosis.

Bone turnover markers such as osteocalcin and C-terminal cross-linked telopeptide are convenient, non-invasive methods for measuring the balance between bone formation and resorption (Vasikaran et al. 2011). In the present study, although the serum level of all markers of bone metabolism (total BMD) were lower in COPD than in asthma patients, we failed to find a significant difference in the specific markers of bone resorption and formation between the two groups. In a study of Xiaomei et al. (2014), circulating biochemical markers of bone resorption and formation were lower in COPD patients than those control subjects, but the mean concentration of vitamin D was similar. We found a weak-to-moderate, statistically significant correlation between all markers of bone metabolism and the patient age in COPD. The lack of this correlation in asthma may be a result of a younger age of asthma patients.

This study has several limitations. First, there were some baseline differences in study groups characteristics that could influenced the results. This particularly refers to the age of patients, disease duration and exposure to tobacco smoke. The median age of asthmatics was 6 years lower than that of COPD patients. Differences in disease duration stem from the fact as asthma often begins at younger age than COPD; the latter is usually related to smoking. Second, the numbers of patients in the study groups were relatively small. This might have affected the statistics or even precluded some statistical analysis in very small sub-groups. Third, definition of the peri- or postmenopausal status was based on the patient age, while the true hormonal status of patients remained uncertain. According to Kaczmarek (2007), the mean age at menopause in Poland is 51.25 years, and those who experience menopause the age of 45 years (5% of the Polish population) are in early menopause. Since 75% of asthma and COPD patients were older than 56.5 and 59.5 years, respectively, we believe we have recruited the truly peri- or postmenopausal women. Finally, the contribution of diet to vitamin D concentration and bone metabolism was not evaluated in this study.

5 Conclusions

Peri- and posmenopausal women with obstructive lung diseases are characterized by decreased serum concentration of vitamin D. Further, vitamin D and total bone mineral density are significantly lower in women suffering from COPD compared to asthma. No relationship between the markers of bone metabolism and bone mineral density was found in peri- and postmenopausal women with obstructive lung diseases. Thus, the rationale to assess these biomarkers remains uncertain. Likewise, the efficacy of vitamin D supplementation in improving pulmonary function, decreasing the frequency of exacerbations, and its impact on the natural course of obstructive lung diseases in peri- and postmenopausal women remain uncertain and warrant further studies.

Conflicts of Interest The authors declare no conflicts of interest in relation to this article.

References

Arshi S, Fallahpour M, Nabavi M, Bemanian MH, Javad-Mousavi SA, Nojomi M, Esmaeilzadeh H, Molatefi R, Rekabi M, Jalali F, Akbarpour N (2014) The effects of Vitamin D supplementation on airway functions in mild to moderate persistent asthma. Ann Allergy Asthma Immunol 113:404–440

Brumpton BM, Langhammer A, Henriksen AH, Camargo CA Jr, Chen Y, Romundstad PR, Mai XM (2016) Vitamin D and lung function decline in adults with asthma: the HUNT Study. Am J Epidemiol 183:739–746

Cassim R, Russell MA, Lodge CJ, Lowe AJ, Koplin JJ, Dharmage SC (2015) The role of circulating 25 hydroxyvitamin D in asthma: a systematic review. Allergy 70:339–354

Confino-Cohen R, Brufman I, Goldberg A, Feldman BS (2014) Vitamin D, asthma prevalence and asthma exacerbations: a large adult population-based study. Allergy 69:1673–1680

Esfandiar N, Alaei F, Fallah S, Babaie D, Sedghi N (2016) Vitamin D deficiency and its impact on asthma severity in asthmatic children. Ital J Pediatr 42:108

GINA (2010) Global initiative for Asthma. Global strategy for Asthma management and prevention updated 2016. http:www.ginasthma.org/2010. Accessed on 23 Nov 2017

GOLD (2010) Global initiative for chronic obstructive lung disease. Global strategy for the diagnosis, management and prevention of COPD. Updated 2017. http:www.goldcopd.org. Accessed on 23 Nov 2017

Graat-Verboom L, Smeenk FW, van den Borne BE, Spruit MA, Jansen FH, van Enschot JW, Wouters EF (2012) Progression of osteoporosis in patients with COPD: a 3-year follow up study. Respir Med 106:861–870

Harlow SD, Gass M, Hall JE, Lobo R, Maki P, Rebar RW, Sherman S, Sluss PM, de Villiers TJ (2012) Executive summary of the stages of reproductive aging workshop +10: addressing the unfinished agenda of staging reproductive aging. J Clin Endocrinol Metab 97:1159–1168

Janssens W, Mathieu C, Boonen S, Decramer M (2011) Vitamin D deficiency and chronic obstructive pulmonary disease: a vicious circle. Vitam Horm 86:379–399

Janssens W, Decramer M, Mathieu C, Korf H (2013) Vitamin D and chronic obstructive pulmonary disease: hype or reality? Lancet Respir Med 1:804–812

Jiao J, King TS, McKenzie M, Bacharier LB, Dixon AE, Codispoti CD, Dunn RM, Grossman NL, Lugogo NL, Ramratnam SK, Traister RS, Wechsler ME, Castro M (2016) Vitamin D3 therapy in patients with asthma complicated by sinonasal disease: secondary analysis of the Vitamin D add-on therapy enhances corticosteroid responsiveness in asthma trial. J Allergy Clin Immunol 1:589–592

Kaczmarek M (2007) Estimation of the age at natural menopause in a population-based study of Polish women. Przegląd Menopauzalny 2:77–82. (Article in polish)

Keating P, Munim A, Hartmann JX (2014) Effect of Vitamin D on T-helper type 9 polarized human memory cells in chronic persistent asthma. Ann Allergy Asthma Immunol 112:154–162

Korn S, Hübner M, Jung M, Blettner M, Buhl R (2013) Severe and uncontrolled adult asthma is associated with vitamin D insufficiency and deficiency. Respir Res 14:15

Larose TL, Langhammer A, Chen Y, Camargo CA Jr, Romundstad P, Mai XM (2015) Serum 25-hydroxyvitamin D levels and lung function in adults with asthma: the HUNT Study. Eur Respir J 45:1019–1026

Li P, Ghazala L, Wright E, Beach J, Morrish D, Vethanayagam D (2015) Prevalence of osteopenia and osteoporosis in patients with moderate to severe asthma in Western Canada. Clin Invest Med 38:23–30

Martineau AR, James WY, Hooper RL, Barnes NC, Jolliffe DA, Greiller CL, Islam K, McLaughlin D, Bhowmik A, Timms PM, Rajakulasingam RK, Rowe M, Venton TR, Choudhury AB, Simcock DE, Wilks M, Degun A, Sadique Z, Monteiro WR, Corrigan CJ, Hawrylowicz CM, Griffiths CJ (2015) Vitamin D3 supplementation in patients with chronic obstructive pulmonary disease (ViDiCO): a multicentre, double blind, randomised controlled trial. Lancet Respir Med 3:120–130

Moberg M, Ringbaek T, Roberts NB, Vestbo J (2014) Association between vitamin D status and COPD phenotypes. Lung 192:493–497

Odler B, Ivancsó I, Somogyi V, Benke K, Tamási L, Gálffy G, Szalay B, Müller V (2015) Vitamin D

deficiency is associated with impaired disease control in asthma-COPD overlap syndrome patients. Int J Chron Obstruct Pulmon Dis 10:2017–2025

Pellegrino R, Viegi G, Brusasco V, Crapo RO, Burgos F, Casaburi R, Coates A, van der Grinten CP, Gustafsson P, Hankinson J, Jensen R, Johnson DC, MacIntyre N, McKay R, Miller MR, Navajas D, Pedersen OF, Wanger J (2005) Interpretative strategies for lung function tests. Eur Respir J 26:948–968

Persson LJ, Aanerud M, Hiemstra PS, Hardie JA, Bakke PS, Eagan TM (2012) Chronic obstructive pulmonary disease is associated with low levels of vitamin D. PLoS One 7:e38934

Płudowski P, Karczmarewicz E, Bayer M et al (2013) Practical guidelines for the supplementation of vitamin D and the treatment of deficits in Central Europe – recommended vitamin D intakes in the general population and groups at risk of vitamin D deficiency. Endokrynol Pol 64:319–327

Schnell K, Weiss CO, Lee T, Krishnan JA, Leff B, Wolff JL, Boyd C (2012) The prevalence of clinically-relevant comorbid conditions in patients with physician-diagnosed COPD: a cross-sectional study using data from NHANES 1999–2008. BMC Pulm Med 12:26

Silva DR, Coelho AC, Dumke A, Valentini JD, de Nunes JN, Stefani CL, da Silva Mendes LF, Knorst MM (2011) Osteoporosis prevalence and associated factors in patients with COPD: a cross-sectional study. Respir Care 56:961–968

Sweeney J, Patterson CC, Menzies-Gow A, Niven RM, Mansur AH, Bucknall C, Chaudhuri R, Price D, Brightling CE, Heaney LG (2016) Comorbidity in severe asthma requiring systemic corticosteroid therapy: cross-sectional data from the optimum patient care research database and the British Thoracic Difficult Asthma Registry. Thorax 71:339–346

Vasikaran S, Eastell R, Bruyère O, Foldes AJ, Garnero P, Griesmacher A, McClung M, Morris HA, Silverman S, Trenti T, Wahl DA, Cooper C, Kanis JA (2011) Markers of bone turnover for the prediction of fracture risk and monitoring of osteoporosis treatment: a need for international reference standards. Osteoporos Int 22:391–420

Veldurthy V, Wei R, Oz L, Dhawan P, Jeon YH, Christakos S (2016) Vitamin D, calcium homeostasis and aging. Bone Res 4:16041

Xiaomei W, Hang X, Lingling L, Xuejun L (2014) Bone metabolism status and associated risk factors in elderly patients with chronic obstructive pulmonary disease (COPD). Cell Biochem Biophys 70:129–134

Zendedel A, Gholami M, Anbari K, Ghanadi K, Bachari EC, Azargon A (2015) Effects of Vitamin D intake on FEV1 and COPD exacerbation: a randomized clinical trial study. Glob J Health Sci 7:243–248

Advs Exp. Medicine, Biology - Neuroscience and Respiration (2018) 39: 37–47
DOI 10.1007/5584_2018_155
© Springer International Publishing AG 2018
Published online: 13 Feb 2018

Oscillations of Subarachnoid Space Width as a Potential Marker of Cerebrospinal Fluid Pulsatility

Marcin Gruszecki, Magdalena K. Nuckowska,
Arkadiusz Szarmach, Marek Radkowski, Dominika Szalewska,
Monika Waskow, Edyta Szurowska, Andrzej F. Frydrychowski,
Urszula Demkow, and Pawel J. Winklewski

Abstract

In the cerebrospinal fluid (CSF) circulation, two components can be distinguished: bulk flow (circulation) and pulsatile flow (back and forth motion). CSF pulsatile flow is generated by both cardiac and respiratory cycles. Recent years have seen increased interest in cardiac- and respiratory-driven CSF pulsatility as an important component of cerebral homeostasis. CSF pulsatility is affected by cerebral arterial inflow and jugular outflow and potentially linked to white matter abnormalities in various diseases, such as multiple sclerosis or hypertension. In this review, we discuss the physiological mechanisms associated with CSF pulsation and its clinical significance. Finally, we explain the concept of using the oscillations of subarachnoid space width as a surrogate for CSF pulsatility.

Keywords

Cerebrospinal fluid · CSF pulsation ·
Oscillations · Pulsatile flow · Subarachnoid
space

M. Gruszecki
Department of Radiology Informatics and Statistics,
Medical University of Gdansk, Gdansk, Poland

M. K. Nuckowska and A. F. Frydrychowski
Department of Human Physiology, Medical University of
Gdansk, Gdansk, Poland

A. Szarmach and E. Szurowska
Second Department of Radiology, Medical University of
Gdansk, Gdansk, Poland

M. Radkowski
Department of Immunopathology of Infectious and
Parasitic Diseases, Warsaw Medical University, Warsaw,
Poland

D. Szalewska
Chair of Rehabilitation Medicine, Medical University of
Gdansk, Gdansk, Poland

M. Waskow
Faculty of Health Sciences, Slupsk Pomeranian
University, Slupsk, Poland

U. Demkow
Department of Laboratory Diagnostics and Clinical
Immunology of Developmental Age, Warsaw Medical
University, Warsaw, Poland

P. J. Winklewski (✉)
Department of Human Physiology, Medical University of
Gdansk, Gdansk, Poland

Second Department of Radiology, Medical University of
Gdansk, Gdansk, Poland

Faculty of Health Sciences, Slupsk Pomeranian
University, Slupsk, Poland
e-mail: pawelwinklewski@wp.pl

1 Introduction

In the cerebrospinal fluid (CSF) circulation, two components can be distinguished: (i) bulk flow (circulation) and (ii) pulsatile flow (back and forth motion). In bulk flow theory, CSF is produced by the choroid plexus and absorbed by arachnoid granulations. The force that drives CSF movement from the ventricular system to the arachnoid granulation for subsequent CSF absorption is generated by a hydrostatic pressure gradient between the site of CSF formation (slightly higher pressure) and its site of absorption (slightly lower pressure). In pulsatile flow theory, movement of the CSF is pulsatile and results from pulsations related to cardiac and respiratory cycles of the subarachnoid portion of the cerebral arteries (Battal et al. 2011).

Recent years have seen increased interest in cardiac- and respiratory-driven CSF pulsatility as an important component of cerebral homeostasis. Bateman et al. (2008) were the first to suggest that CSF pulsatility may represent a key mechanism of absorbing energy stored in heart-generated arterial waves. Changes in the sophisticated Windkessel properties of CSF pulsatility (Fig. 1) might be associated with micro-injuries within brain white matter (Jolly et al. 2013). That has been confirmed by Beggs et al. (2016b) who demonstrated that dirty-appearing white matter is associated with altered CSF pulsatility in patients suffering from hypertension.

We have previously shown, using methodology based on infrared light, that jugular flow impairment is associated with increased CSF pulsatility (Frydrychowski et al. 2012b). This finding was confirmed 2 years later using cine phase contrast (cine) magnetic resonance imaging (MRI) (Beggs et al. 2014). Augmented CSF pulsatility, a consequence of impaired jugular outflow, is in perfect agreement with the CSF Windkessel model above outlined. Energy stored in the form of pulse pressure, and manifested as a pulsatile flow, must be dampened. This process occurs by shifting CSF and venous blood, i.e., the Windkessel effect. Therefore, pulse dampening might be disrupted by either too large input (large arterial pulsations) or reduced compliance in output (impaired venous outflow).

In this review, we first discuss the physiological mechanisms associated with CSF pulsation, including two main drivers, i.e., cardiac and respiration cycles. Secondly, we present the state of the art data regarding the clinical significance of proper CSF pulsatility. Thirdly, we present near infrared transillumination back scattering sounding (NIR-T/BSS), a method developed by our team to measure a surrogate of CSF pulsatility, namely subarachnoid space width oscillation, including recent research data. Finally, the future directions of this quickly developing field of medicine are discussed.

2 Physiological Mechanisms Defining Cerebrospinal Fluid Pulsatility

DuBoulay (1966) and DuBoulay et al. (1972), measuring CSF pulsatility during pneumoencephalography, myelography and ventriculography, have reported that pulsations in the normal cervical subarachnoid space (SAS) paralleled arterial pulsations, and that these CSF pulsations diminished below the upper thoracic region. They concluded that CSF flow in the basal cisterns is a consequence of the rhythmic expansion and contraction of the brain during systole and diastole. Consequently, the authors argue that the causes of CSF movement were ultimately related to the pressure and elasticity provided by the arteries and veins. Lin and Kricheff (1972) have demonstrated, in healthy subjects, caudal flows in cardiac systole and cranial flows in cardiac diastole, while in subjects with obstruction of the spinal CSF pathways, a lower pulsation amplitude was observed above and below the block.

Quencer et al. (1990) have described normal patterns of pulsatile flow in the ventricles, cisterns, and cervical SAS, using the cine MRI technique with cardiac gating. In cardiac systole, there is a downward (caudal) flow of CSF in the aqueduct of Sylvius, the foramen of Magendie, the basal cisterns, and the dorsal and ventral SAS, while during diastole, upward (cranial) flow of

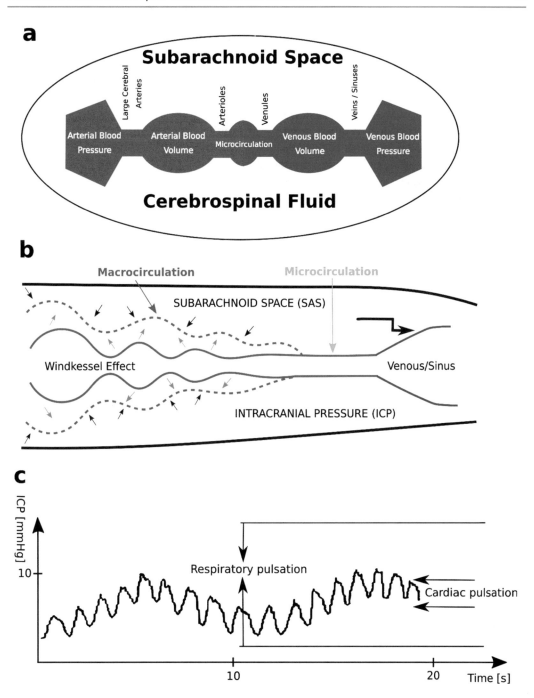

Fig. 1 Arterial and CSF pulsation through the foramen magnum are so closely coupled that CSF oscillations are in a state of resonance with arterial pulsations (Egnor et al. 2002). The beneficial effect of this resonance is that a dynamic oscillator so configured will maximally dampen the arterial pulsations before their reaching the capillary bed by removing pulsation energy in systole and reapplying it to the arterial tree in diastole (Bateman et al. 2008). The capillary bed of the brain requires non-pulsatile continuous flow, so the effective dampening of arterial pulsatility is essential

CSF in the same structures is observed. Importantly, the relationship between the cardiac cycle and CSF pulsations has been demonstrated in both magnitude and phase reconstructions of MRI images. For the first time, calculations of the actual CSF velocity have been obtained from the phase reconstruction images. Under conditions that result in alterations in flow, cine MRI is capable of showing either obstruction or excessively turbulent flow within CSF pathways. These early results have been confirmed by several research teams (Wagshul et al. 2006; Greitz et al. 1992; Schroth and Klose 1992a; Enzmann and Pelc 1991).

A comparative MRI analysis of signal intensity and the proton phase in different CSF compartments led to the conclusion that cardiac-related pulsatile CSF flow begins with rapid, early systolic displacement of a relatively large volume of CSF from the intracranial SAS into the cervical spinal canal. The heart-generated increased systolic intracranial blood volume accumulates mainly in the richly vascularized compartments of the brain. Consequently, CSF is initially displaced from the SAS adjacent to the richly vascularized grey matter into the cervical spinal canal immediately after the inflow of systolic blood (Schroth and Klose 1992a).

As early as in 1886, it has been shown that intracranial pressure fluctuations in the rabbit cisterna magna depend not only on the heart cycle, but also on the respiration cycle (Knoll 1886). Expiration was accompanied by an increase in intracranial pressure, while its decline was observed during inspiration. With the introduction of spinal puncture into clinical practice at the end of the nineteenth century, the dependence of intracranial pressure fluctuations on the heart and respiratory cycles was confirmed in humans (Becher 1924; Becher 1919).

Cardiac-related pulsations of the CSF can be investigated using ECG-gated MRI. However, real-time detection of respiratory movement is impossible using this technique. The limitation can be bypassed if no resolution of the second direction in the plane is necessary and no phase encoding gradient is used, resulting in a reduction in the investigation time to 20 ms (Schroth and Klose 1992b). Therefore, the real-time acquisition and evaluation (RACE) technique enables the evaluation of respiratory movements and CSF pulsation over an extended time period. However, due to technical restrictions, such as a local resolution in only one plane and difficulties in identifying the phase shift of CSF protons, quantitative studies and exact analysis of the time course of CSF flow with respect to the heart beat and respiration are difficult.

Nevertheless, the RACE technique has helped demonstrate that caudal CSF flow in the anterior cervical SAS dominates during inspiration, whereas an increase in cephalad flow occurs during expiration. Caudal flow acceleration occurs immediately upon the beginning of inspiration, but ceases with abdominal compression. In the aqueduct, cardiac-related CSF pulsation is affected in a similar way by respiration. The systolic pulsatile flow component from the third to the fourth ventricle downward increases during inspiration after a delay of two to three cardiac cycles, whereas during the late phase of expiration, backflow from the fourth to the third ventricle accelerates (Schroth and Klose 1992b). A physiological explanation for the observed CSF motion is that the spinal epidural veins are emptied due to the inspiratory negative pressure in the thoracic cavity; this leads to a rapid caudal acceleration of CSF flow in the spinal canal. In addition, diminished thoracic pressure directly affects the hydrostatic pressure that drives the low-resistance paravenous, venous, and lymphatic CSF drainage (Chen et al. 2015; Dreha-Kulaczewski et al. 2015).

To summarize, CSF pulsatility is affected by both cardiac and respiratory cycles. However, due to a limited temporal resolution of MRI, it is not possible to assess the detailed relationship between these two processes.

3   Clinical Significance of Cerebrospinal Fluid Pulsatility

Zamboni et al. (2009b) have proposed a causal link between multiple sclerosis and the altered modality of venous return, determined by multiple extracranial venous strictures, known as chronic cerebrospinal venous insufficiency. In another publication, the same investigators implied that chronic cerebrospinal venous insufficiency causes venous reflux, leading to iron buildup in the brain, which they suggest is a primary event in the pathogenesis of multiple sclerosis, as it triggers subsequent inflammatory injury to the central nervous system (Singh and Zamboni 2009). Actually, the potential role of increased iron deposition and iron-mediated injury to the central nervous system is not exclusive to multiple sclerosis and is well-documented in many neurological disorders, particularly neurodegenerative diseases (Benarroch 2009).

Zamboni et al. (2009a) have described the results of an open-label, unblinded trial of venous angioplasty in 65 patients with multiple sclerosis. Although the concept of performing venous angioplasty or stenting to treat multiple sclerosis was highly criticized (Siddiqui et al. 2014), including a Federal Drug Administration warning related to "controversial experimental procedure for treating multiple sclerosis" (Kuehn 2012), it reinvigorated interest in the CSF pulsatility and the role of venous outflow in regulating pulsatile CSF flow. In another study, Zamboni et al. (2009c) have used Doppler ultrasound imaging to demonstrate the relationship between the severity of cerebral venous hemodynamic insufficiency and distorted CSF flow dynamics in 16 patients presenting with relapsing-remitting multiple sclerosis and in eight healthy control subjects. Unfortunately, the Doppler ultrasound-based criteria to diagnose cerebral venous hemodynamic insufficiency in this study are considered insufficient, so that this hypothesis remains controversial (Caprio et al. 2017; Laukontaus et al. 2017).

Nevertheless, 2 years later, Zivadinov et al. (2011) have used pre- and post-contrast susceptibility-weighted imaging venography to show that multiple sclerosis patients present with a higher number of venous stenosis, indicative of severe chronic cerebrospinal venous insufficiency and decreased venous vasculature in brain parenchyma. The authors concluded that reduced metabolism and morphological changes to the venous vasculature might take place in patients with multiple sclerosis. Quite interestingly, augmented CSF pulsatility seems associated with an increased internal jugular vein cross-sectional area in the lower cervix, independent of age and cardiovascular risk factors, which suggests the presence of a biomechanical link between the two. This relationship is altered in multiple sclerosis patients (Beggs et al. 2016a). Further, percutaneous transluminal angioplasty in patients with multiple sclerosis and chronic cerebrospinal venous insufficiency increases CSF flow and decreases CSF velocity, which are indicative of improved venous parenchyma drainage (Zivadinov et al. 2013). These findings have been obtained using a widely accepted MRI technique, and validated in multiple sclerosis patients (Beggs et al. 2012; Magnano et al. 2012).

Increased CSF pulsatility has also been observed in individuals with impaired jugular outflow, otherwise remaining healthy (Beggs et al. 2014). Likewise, healthy aging seems associated with the enlargement of internal jugular vein cross-sectional area (Magnano et al. 2016). Sex differences have been reported in the relationship between brain volume and internal jugular vein cross-sectional area in healthy individuals without a neurologic disease. After the adjustment for age, the association between the normalized whole brain or grey matter volume and the internal jugular vein cross-sectional area is positive only in males (Belov et al. 2017). In patients with Alzheimer's disease, jugular venous reflux might be linked to white matter abnormalities (Chung et al. 2014). Finally, it has been shown that leukoaraiosis might be associated with increased CSF pulsatility linked to arterial hypertension (Beggs et al. 2016b). Increased arterial pulsatility may augment perivascular shear stress and lead to accumulated damage to perivascular oligodendrocytes, resulting in microstructural changes in white matter and contributing to leukoaraiosis over time (Jolly et al. 2013). The link between hypertension, CSF pulsatility, and white matter changes, if confirmed, may actually suggest that studying CSF pulsatility abnormalities is not a niche research topic, but rather a pathological phenomenon encompassing a large patient population.

To summarize, CSF pulsatility is affected by cerebral arterial inflow and jugular outflow, and it is potentially linked to white matter abnormalities in various diseases.

4   Oscillation of Subarachnoid Space Width

In the last decade, a new method based on infrared radiation, called near-infrared transillumination/backscattering sounding (NIR-T/BSS), has been developed. NIR-T/BSS allows for the measurement of SAS width to determine changes in CSF volume (Frydrychowski et al. 2011; Frydrychowski and Pluciński 2007; Pluciński and Frydrychowski 2007). Contrary to near-infrared spectroscopy, which relies on the absorption of infrared light by hemoglobin (Ferrari and Quaresima 2012), NIR-T/BSS uses the SAS filled with translucent CSF as a propagation duct for infrared radiation (Frydrychowski et al. 2002; Pluciński et al. 2000). The signal received by the distal detector is divided over the signal received by the proximal detector. This reduces the proportional factors that affect each of the two

signals in an identical way, due to the fact that the quotient of these factors assumes a value of 1. Both the dividend, i.e., the power of the distal signal, and the divisor, i.e., the power of the proximal signal, are influenced by the width of the SAS and by any factor capable of changing that width. Therefore, the quotient of the two signals, hereafter called the transillumination quotient (TQ), is sensitive to changes in the width of the SAS. The oscillations in TQ have their origin in different modulations of signals from proximal and distal detectors, namely in the modulation of the distal signal on its way through the SAS. This happens because only the distal detector receives the radiation propagated within the SAS. The propagation of infrared light in the skin and bone is hindered compared to the clear, translucent CSF, and with the distal detector placed far enough from the signal emitter, no radiation propagated in the superficial tissue layers can reach the distal detector (Frydrychowski et al. 2002; Pluciński et al. 2000). The power of the infrared light stream reaching the distal detector is directly proportional to the width of the SAS (Fig. 2). The wider the SAS or the propagation duct, the more radiation reaches the distal detector and the

a

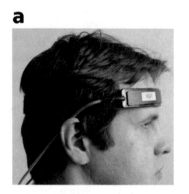

b

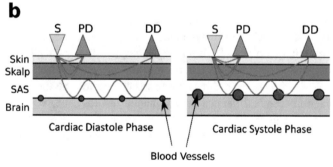

Fig. 2 Schematic localization of the near-infrared transillumination/backscattering sounding (NIR-T/BSS) sensor unit, spatial relations between its elements and the paths followed by radiation in the head tissues between the signal emitter (S) and the proximal and distal detectors (PD and DD, respectively). The heart-generated systolic intracranial blood volume is the most accentuated in the richly vascularized compartments of the brain. Consequently, during systole, CSF is displaced from the adjacent subarachnoid space (SAS) to the richly vascularized grey matter into the cervical spinal canal (Schroth and Klose 1992a), leading to a reduction in the SAS width. The opposite happens during diastole (the individual shown in the picture consented to the use of the image)

greater the signal from that detector, which is the dividend in the calculation of the TQ.

Using empirical or non-linear mode decomposition methods, several components of the width signal of SAS can be subtracted. The power spectrum density of SAS oscillations shows clear peaks at the cardiac and respiratory frequencies. Therefore, NIR-T/BSS allows for simultaneous assessment of cardiac- and respiratory-mediated changes in SAS width (Fig. 3), a modality not offered by MRI. Further, high sampling frequency (70 Hz) of NIR-T/BSS allows for signal analysis up to 35 Hz according to the Nyquist theorem. Thus, using NIR-T/BSS, it is possible to accurately register rapid SAS oscillations secondary to systolic diastolic changes in blood volume of cerebral circulation (Kalicka et al. 2017).

Calculating wavelet phase and amplitude coherences, one may identify similarities between the two oscillators. Combining NIR-T/BSS recordings and mathematical modeling, a proportion of systemic blood pressure oscillations with corresponding heart-driven SAS changes may be precisely identified.

To assess the relationship between blood pressure and SAS oscillations, we used wavelet transform analysis, as it ensures windows of adjustable lengths, thereby providing the benefit of showing high resolution at cardiac frequency. Compared with autoregressive estimation, wavelet transform is calculated directly from the data, and the limitations of linear modeling and the choice of model order are thus avoided (Bernjak et al. 2012; Shiogai et al. 2010; Stefanovska et al. 1999).

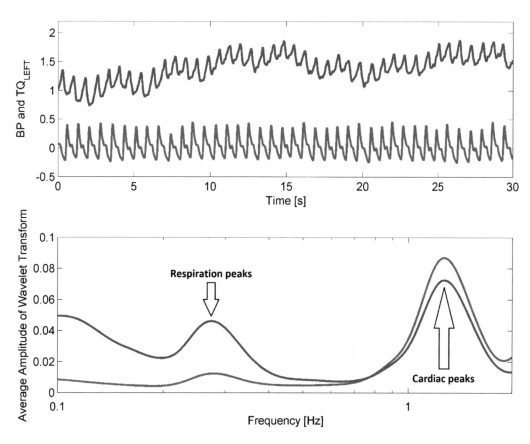

Fig. 3 Representative 2-min tracings of baseline signal. The subarachnoid space (SAS) signal (upper panel in blue) is less regular than the blood pressure signal (upper panel in red). Wavelet transform analysis reveals blood pressure and SAS peaks at cardiac and respiratory frequencies. *BP* blood pressure, *TQ* transillumination quotient

Coupling functions enabled us to unveil a new perspective on how neurophysiological mechanisms are affected by various stimuli. In particular, we have shown that the coherence of blood pressure and SAS amplitudes decreases during apneic episodes in normal healthy subjects (Winklewski et al. 2015b), but it remains stable during such episodes in professional divers (Winklewski et al. 2015a). Quite surprisingly, the sympathetic nervous system appears to stabilize the above-mentioned relationship (Winklewski et al. 2015c). On the contrary, changes in hemoglobin oxygen saturation (i.e., hypoxia) and intrathoracic pressure evoke large swings in the coherence of blood pressure and SAS amplitude (Wszedybyl-Winklewska et al. 2017a, b, c). Therefore, the stimuli like apnea, hypoxia, and increased respiratory resistance may increase CSF pulsation instability.

Importantly, NIR-T/BSS and MRI are comparable and give equivalent modalities in the measurement of SAS width. Correlation and regression analysis of changes in SAS width in the supine and prone positions measured with MRI and NIRT-B/SS demonstrate a high interdependence between both methods (r = 0.81, p < 0.001; Frydrychowski et al. 2012a). Moreover, jugular vein compression produces a marked increase in pulsation of a cardiac component of SAS width (Fig. 4). Increases in velocity of cerebral blood flow and in pulsation of SAS cardiac component, observed during acute impairment of venous outflow, are consistent with the model proposed by Bateman et al. (2008). Our finding that jugular vein congestion may result in augmented SAS pulsatility at cardiac frequency has been confirmed Beggs et al. (2014), who report elevated CSF pulsatility in the aqueduct of Sylvius.

To summarize, SAS pulsatility is directly affected by jugular outflow, while changes in the relationship between blood pressure and SAS width oscillations may reflect the overall increase in instability with respect to CSF pulsatility. The possible limitation of the research above reviewed, which investigates the interplay between jugular outflow and CSF pulsatility, is

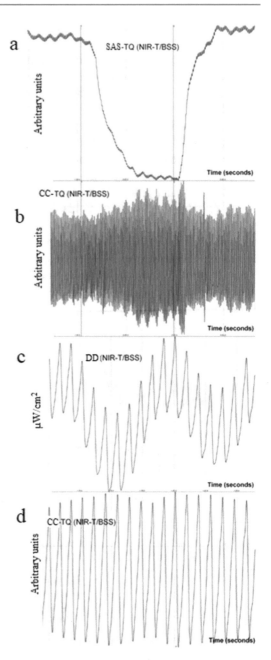

Fig. 4 Effect of acute bilateral jugular vein compression on NIR-T/BSS variables. (**a**) SAS-TQ after jugular vein compression. A decline in SAS width is visible; (**b**) CC-TQ after jugular vein compression. An increase in pulsation of SAS cardiac component is evident; (**c**) and (**d**) retraced signals from the distant detector (DD) and from cardiac SAS component (CC-TQ), respectively. The absence of the 'cutting' effect and sharp edges indicates that the SAS width, although narrowed, is still wide enough for proper signal registration. $\mu W/cm^2$ – microwatt/cm^2 (Reprinted from Frydrychowski et al. 2012b; Creative Commons Attribution License)

that it has been based on single-site data. Therefore, the replication of these findings is required.

---

5   Perspectives

CSF pulsatility assessment may provide new insights into the understanding of the pathomechanisms leading to vascular dementia and neurodegenerative diseases. To further develop this promising area of research, there is a need for better tools, including MRI methods with higher temporal resolution and augmented flexibility, to simultaneously trace cardiac and respiratory cycles. The near infrared transillumination back scattering sounding may offer several advantages in terms of multi-signal analysis, assessment of coupling functions, and advanced mathematical modeling. Further validation of the method is needed, such as a direct comparison between the width of the cervical subarachnoid space and the intracranial pressure signals in intensive care unit patients. Studies including the near infrared transillumination back scattering sounding and the well-recognized radiological imaging methods, like MRI or computed tomography, are warranted.

Competing Interests Drs. Andrzej Frydrychowski and Pawel J. Winklewski are stakeholders in NIRTI SA, Wierzbice, Poland, which has developed and produces the NIR-T/BSS device. NIRTI SA had no role in manuscript design, preparation, or the decision to publish.

References

Bateman GA, Levi CR, Schofield P, Wang Y, Lovett EC (2008) The venous manifestations of pulse wave encephalopathy: windkessel dysfunction in normal aging and senile dementia. Neuroradiology 50:491–497

Battal B, Kocaoglu M, Bulakbasi N, Husmen G, Tuba Sanal H, Tayfun C (2011) Cerebrospinal fluid flow imaging by using phase-contrast MR technique. Br J Radiol 84:758–765

Becher E (1919) Beobachtungen ueber die Abhaengigkeit des Lumbaldruckes yon der Kopfhaltung. Dtsch Z Nervenheilkd 63:89–96 (Article in German)

Becher E (1924) Ueber Druckverhaeltnisse im Liqour cerebrospinalis. Grenzgeb Med Chir 35:324–332 (Article in German)

Beggs CB, Magnano C, Belov P, Krawiecki J, Ramasamy DP, Hagemeier J, Zivadinov R (2016a) Internal jugular vein cross-sectional area and cerebrospinal fluid pulsatility in the aqueduct of sylvius: a comparative study between healthy subjects and multiple sclerosis patients. PLoS One 11:e0153960

Beggs CB, Magnano C, Shepherd SJ, Belov P, Ramasamy DP, Hagemeier J, Zivadinov R (2016b) Dirty-appearing white matter in the brain is associated with altered cerebrospinal fluid pulsatility and hypertension in individuals without neurologic disease. J Neuroimaging 26:136–143

Beggs CB, Magnano C, Shepherd SJ, Marr K, Valnarov V, Hojnacki D, Bergsland N, Belov P, Grisafi S, Dwyer MG, Carl E, Weinstock-Guttman B, Zivadinov R (2014) Aqueductal cerebrospinal fluid pulsatility in healthy individuals is affected by impaired cerebral venous outflow. J Magn Reson Imaging 40:1215–1222

Beggs CB, Shepherd SJ, Dwyer MG, Polak P, Magnano C, Carl E, Poloni GU, Weinstock-Guttman B, Zivadinov R (2012) Sensitivity and specificity of SWI venography for detection of cerebral venous alterations in multiple sclerosis. Neurol Res 34:793–801

Belov P, Magnano C, Krawiecki J, Hagemeier J, Bergsland N, Beggs C, Zivadinov R (2017) Age-related brain atrophy may be mitigated by internal jugular vein enlargement in male individuals without neurologic disease. Phlebology 32:125–134

Benarroch EE (2009) Brain iron homeostasis and neurodegenerative disease. Neurology 72:1436–1440

Bernjak A, Stefanovska A, McClintock PVE, Owen-Lynch PJ, Clarkson PBM (2012) Coherence between fluctuations in blood flow and oxygen saturation. Fluct Noise Lett 11:1240013

Caprio MG, Marr K, Gandhi S, Jakimovski D, Hagemeier J, Weinstock-Guttman B, Zivadinov R, Mancini M (2017) Centralized and local color Doppler ultrasound reading agreement for diagnosis of the chronic cerebrospinal venous insufficiency in patients with Multiple Sclerosis. Curr Neurovasc Res 14:266–273

Chen L, Beckett A, Verma A, Feinberg DA (2015) Dynamics of respiratory and cardiac CSF motion revealed with real-time simultaneous multi-slice EPI velocity phase contrast imaging. NeuroImage 122:281–287

Chung CP, Beggs C, Wang PN, Bergsland N, Shepherd S, Cheng CY, Ramasamy DP, Dwyer MG, Hu HH, Zivadinov R (2014) Jugular venous reflux and white matter abnormalities in Alzheimer's disease: a pilot study. J Alzheimers Dis 39:601–609

Dreha-Kulaczewski S, Joseph AA, Merboldt KD, Ludwig HC, GaÈrtner J, Frahm J (2015) Inspiration is the major regulator of human CSF flow. J Neurosci 35:2485–2491

DuBoulay GH (1966) Pulsatile movements in the CSF pathways. Br J Radiol 139:255–262

DuBoulay GH, O'Connell J, Currie J, Bostic KT, Verity P (1972) Further investigations on pulsatile movements in the cerebrospinal fluid pathways. Acta Radiol 113:496–523

Egnor M, Zheng L, Rosiello A, Gutman F, Davis R (2002) A model of pulsations in communicating hydrocephalus. Pediatr Neurosurg 36:281–303

Enzmann DR, Pelc NJ (1991) Normal flow patterns of intracranial and spinal cerebrospinal fluid defined with phase-contrast cine MR imaging. Radiology 178:467–474

Ferrari M, Quaresima V (2012) A brief review on the history of human functional near-infrared spectroscopy (fNIRS) development and fields of application. NeuroImage 63:921e35

Frydrychowski AF, Gumiński W, Rojewski M, Kaczmarek J, Juzwa W (2002) Technical foundations for noninvasive assessment of changes in the width of the subarachnoid space with near-infrared transillumination-backscattering sounding (NIR-TBSS). IEEE Trans Biomed Eng 49:887–904

Frydrychowski AF, Pluciński J (2007) New aspects in assessment of changes in width of subarachnoid space with near-infrared transillumination-backscattering sounding, part 2: clinical verification in the patient. J Biomed Opt 12:044016

Frydrychowski AF, Szarmach A, Czaplewski B, Winklewski PJ (2012a) Subarachnoid space: new tricks by an old dog. PLoS One 7:e37529

Frydrychowski AF, Winklewski PJ, Guminski W (2012b) Influence of acute jugular vein compression on the cerebral blood flow velocity, pial artery pulsation and width of subarachnoid space in humans. PLoS One 7: e48245

Frydrychowski AF, Wszedybyl-Winklewska M, Guminski W, Przyborska A, Kaczmarek J, Winklewski PJ (2011) Use of near infrared transillumination/back scattering sounding (NIR-T/BSS) to assess effects of elevated intracranial pressure on width of subarachnoid space and cerebrovascular pulsation in animals. Acta Neurobiol Exp 71:313–321

Greitz D, Wirestam R, Franck A, Nordell B, Thomsen C, Ståhlberg F (1992) Pulsatile brain movement and associated hydrodynamics studied by magnetic resonance phase imaging. The Monro-Kellie doctrine revisited. Neuroradiology 34:370–380

Jolly TA, Bateman GA, Levi CR, Parsons MW, Michie PT, Karayanidis F (2013) Early detection of microstructural white matter changes associated with arterial pulsatility. Front Hum Neurosci 7:782

Kalicka R, Mazur K, Wolf J, Frydrychowski AF, Narkiewicz K, Winklewski PJ (2017) Modelling of subarachnoid space width changes in apnoea resulting as a function of blood flow parameters. Microvasc Res 113:16–21

Knoll P (1886) Ueber die Druckschwankungen in der Cerebrospinalfluessigkeit und den Wechsel in der Blutfuelle des centralen Nervensystems. Sitzungs-berichte der Wiener kaiserlichen Akademie der Wissenschaften. Bd. XCIII. Heft 5. Abtheilung 3. Jahrg. 1886. Wien 1886 (Article in German)

Kuehn BM (2012) FDA warns about the risks of unproven surgical therapy for multiple sclerosis. JAMA 307:2575–2576

Laukontaus SJ, Pekkola J, Numminen J, Kagayama T, Lepäntalo M, Färkkilä M, Atula S, Tienari P, Venermo M (2017) Magnetic resonance imaging of internal jugular veins in multiple sclerosis: interobserver agreement and comparison with Doppler ultrasound examination. Ann Vasc Surg 42:84–92

Lin JP, Kricheff II (1972) Angiographic investigation of cerebral aneurysms. Technical aspects. Radiology 105 (1):69–76

Magnano C, Belov P, Krawiecki J, Hagemeier J, Beggs C, Zivadinov R (2016) Internal jugular vein cross-sectional area enlargement is associated with aging in healthy individuals. PLoS One 11:e0149532

Magnano C, Schirda C, Weinstock-Guttman B, Wack DS, Lindzen E, Hojnacki D, Bergsland N, Kennedy C, Belov P, Dwyer MG, Poloni GU, Beggs CB, Zivadinov R (2012) Cine cerebrospinal fluid imaging in multiple sclerosis. J Magn Reson Imaging 36:825–834

Pluciński J, Frydrychowski AF (2007) New aspects in assessment of changes in width of subarachnoid space with near-infrared transillumination/backscattering sounding, part 1: Monte Carlo numerical modeling. J Biomed Opt 12:044015

Pluciński J, Frydrychowski AF, Kaczmarek J, Juzwa W (2000) Theoretical foundations for noninvasive measurement of variations in the width of the subarachnoid space. J Biomed Opt 5:291–299

Quencer RM, Post MJ, Hinks RS (1990) Cine MR in the evaluation of normal and abnormal CSF flow: intracranial and intraspinal studies. Neuroradiology 32:371–391

Schroth G, Klose U (1992a) Cerebrospinal fluid flow. I. Physiology of cardiac-related pulsation. Neuroradiology 35:1–9

Schroth G, Klose U (1992b) Cerebrospinal fluid flow. I. Physiology of respiration-related pulsation. Neuroradiology 35:10–15

Siddiqui AH, Zivadinov R, Benedict RH, Karmon Y, Yu J, Hartney ML, Marr KL, Valnarov V, Kennedy CL, Ramanathan M, Ramasamy DP, Dolic K, Hojnacki DW, Carl E, Levy EI, Hopkins LN, Weinstock-Guttman B (2014) Prospective randomized trial of venous angioplasty in MS (PREMiSe). Neurology 83:441–449

Shiogai Y, Stefanovska A, McClintock PV (2010) Nonlinear dynamics of cardiovascular ageing. Phys Rep 488:51–110

Singh AV, Zamboni P (2009) Anomalous venous blood flow and iron deposition in multiple sclerosis. J Cereb Blood Flow Metab 29:1867–1878

Stefanovska A, Bracic M, Kvernmo HD (1999) Wavelet analysis of oscillations in the peripheral blood circulation measured by laser Doppler technique. IEEE Trans Biomed Eng 46:1230–1239

Wagshul ME, Chen JJ, Egnor MR, McCormack EJ, Roche PE (2006) Amplitude and phase of cerebrospinal fluid pulsations: experimental studies and review of the literature. J Neurosurg 104:810–819

Winklewski PJ, Barak O, Madden D, Gruszecka A, Gruszecki M, Guminski W, Kot J, Frydrychowski AF, Drvis I, Dujic Z (2015a) Effect of maximal apnoea easy-going and struggle phases on subarachnoid width and pial artery pulsation in elite breath-hold divers. PLoS One 10:e0135429

Winklewski PJ, Gruszecki M, Wolf J, Swierblewska E, Kunicka K, Wszedybyl-Winklewska M, Guminski W, Zabulewicz J, Frydrychowski AF, Bieniaszewski L, Narkiewicz K (2015b) Wavelet transform analysis to assess oscillations in pial artery pulsation at the human cardiac frequency. Microvasc Res 99:86–91

Winklewski PJ, Tkachenko Y, Mazur K, Kot J, Gruszecki M, Guminski W, Czuszynski K, Wtorek J, Frydrychowski AF (2015c) Sympathetic activation does not affect the cardiac and respiratory contribution to the relationship between blood pressure and pial artery pulsation oscillations in healthy subjects. PLoS One 10:e0135751

Wszedybyl-Winklewska M, Wolf J, Swierblewska E, Kunicka K, Gruszecka A, Gruszecki M, Kucharska W, Winklewski PJ, Zabulewicz J, Guminski W, Pietrewicz M, Frydrychowski AF, Bieniaszewski L, Narkiewicz K (2017a) Acute hypoxia diminishes the relationship between blood pressure and subarachnoid space width oscillations at the human cardiac frequency. PLoS One 12:e0172842

Wszedybyl-Winklewska M, Wolf J, Swierblewska E, Kunicka K, Mazur K, Gruszecki M, Winklewski PJ, Frydrychowski AF, Bieniaszewski L, Narkiewicz K (2017b) Increased inspiratory resistance affects the dynamic relationship between blood pressure changes and subarachnoid space width oscillations. PLoS One 12:e0179503

Wszedybyl-Winklewska M, Wolf J, Szarmach A, Winklewski PJ, Szurowska E, Narkiewicz K (2017c) Central sympathetic nervous system reinforcement in obstructive sleep apnoea. Sleep Med Rev. https://doi.org/10.1016/j.smrv.2017.08.006

Zamboni P, Galeotti R, Menegatti E, Malagoni AM, Gianesini S, Bartolomei I, Mascoli F, Salvi F (2009a) A prospective open-label study of endovascular treatment of chronic cerebrospinal venous insufficiency. J Vasc Surg 50:1348–1358

Zamboni P, Galeotti R, Menegatti E, Malagoni AM, Tacconi G, Dall'Ara S, Bartolomei I, Salvi F (2009b) Chronic cerebrospinal venous insufficiency in patients with multiple sclerosis. J Neurol Neurosurg Psychiatry 80:392–399

Zamboni P, Menegatti E, Weinstock-Guttman B, Schirda C, Cox JL, Malagoni AM, Hojanacki D, Kennedy C, Carl E, Dwyer MG, Bergsland N, Galeotti R, Hussein S, Bartolomei I, Salvi F, Zivadinov R (2009c) The severity of chronic cerebrospinal venous insufficiency in patients with multiple sclerosis is related to altered cerebrospinal fluid dynamics. Funct Neurol 24:133–138

Zivadinov R, Magnano C, Galeotti R, Schirda C, Menegatti E, Weinstock-Guttman B, Marr K, Bartolomei I, Hagemeier J, Malagoni AM, Hojnacki D, Kennedy C, Carl E, Beggs C, Salvi F, Zamboni P (2013) Changes of cine cerebrospinal fluid dynamics in patients with multiple sclerosis treated with percutaneous transluminal angioplasty: a case-control study. J Vasc Interv Radiol 24:829–838

Zivadinov R, Ramanathan M, Dolic K, Marr K, Karmon Y, Siddiqui AH, Benedict RH, Weinstock-Guttman B (2011) Chronic cerebrospinal venous insufficiency in multiple sclerosis: diagnostic, pathogenetic, clinical and treatment perspectives. Expert Rev Neurother 11:1277–1294

Advs Exp. Medicine, Biology - Neuroscience and Respiration (2018) 39: 49–70
DOI 10.1007/5584_2018_154
© Springer International Publishing AG 2018
Published online: 11 Feb 2018

Very High Frequency Oscillations of Heart Rate Variability in Healthy Humans and in Patients with Cardiovascular Autonomic Neuropathy

Mario Estévez-Báez, Calixto Machado, Julio Montes-Brown, Javier Jas-García, Gerry Leisman, Adam Schiavi, Andrés Machado-García, Claudia Carricarte-Naranjo, and Eli Carmeli

Abstract

Literature reports on the very high frequency (VHF) range of 0.4–0.9 Hz in heart rate variability (HRV) are scanty. The VHF presence in cardiac transplant patients and other conditions associated with reduced vagal influence on the heart encouraged us to explore this spectral band in healthy subjects and in patients diagnosed with cardiac autonomic neuropathy (CAN), and to assess the potential clinical value of some VHF indices. The study included 80 healthy controls and 48 patients with spinocerebellar ataxia type 2 (SCA2) with CAN. The electrocardiographic recordings of short 5-min duration were submitted to three different spectral analysis methods, including the most generally accepted procedure, and the two novel methods using the Hilbert-Huang transform. We demonstrated the presence of VHF activity in both groups of subjects.

M. Estévez-Báez and C. Machado
Institute of Neurology and Neurosurgery, Havana, Cuba

J. Montes-Brown
General Hospital 'Ernesto Guevara de la Serna', Las Tunas, Cuba

Latin American School of Medicine, Havana, Cuba

J. Jas-García
Center for Sports Research, Havana, Cuba

G. Leisman (✉)
Faculty of Social Welfare and Health Science, University of Haifa, Haifa, Israel

National Institute for Brain and Rehabilitation Sciences, Nazareth, Israel

University of the Medical Sciences, 'Manuel Fajardo' Medical School, Havana, Cuba
e-mail: g.leisman@alumni.manchester.ac.uk

A. Schiavi
Anesthesiology and Critical Care Medicine, Neurosciences Critical Care Division, Johns Hopkins Hospital, Baltimore, MD, USA

A. Machado-García and C. Carricarte-Naranjo
Department of Biology, University of Havana, Havana, Cuba

E. Carmeli
Physical Therapy Department, Faculty of Social Welfare and Health Science, University of Haifa, Haifa, Israel

However, VHF power spectral density, expressed in relative normalized units, was significantly greater in the SCA2 patients than that in healthy subjects, amounting to $36.1 \pm 17.4\%$ vs. $22.9 \pm 14.1\%$, respectively, as also was the instantaneous VHF spectral frequency, 0.58 ± 0.05 vs. 0.64 ± 0.07 Hz, respectively. These findings were related to the severity of CAN. We conclude that VHF activity of HRV is integral to the cardiovascular autonomic control.

Keywords

Autonomic nervous system · Cardiovascular autonomic neuropathy · Denervation · Empirical mode decomposition · Heart rate variability · Power spectrum analysis · Spinocerebellar ataxia

1   Introduction

Several methods have been successfully introduced to study heart rate variability (HRV) during the last 50 years (Task Force of ESC and NASPE 1996; Sassi et al. 2015). Spectral frequency content of HRV has been the subject of extensive investigation. Different spectral components from short-term electrocardiogram (ECG) recordings (2–10 min) have been identified, and associated with a variety of physiologic mechanisms (Estévez-Báez et al. 2016a, b; Machado et al. 2014; Cygankiewicz and Zareba 2013; Kuusela 2013; Machado-Ferrer et al. 2013; Vinik and Erbas 2013; MacFarlane et al. 2010). The upper limit of HRV frequency is widely accepted to be no higher than 0.4 Hz and corresponds to the HRV band of high frequencies range of 0.15–0.40 Hz (Task Force of ESC and NASPE 1996). Nevertheless, recent studies demonstrate the existence of very high frequency (VHF) oscillations, in a range of 0.4–0.9 Hz, which have been unraveled in 15 healthy subjects with the use of pulse rate variability, a surrogate of HRV, during non-stationary conditions (Chang et al. 2014a, b). The VHF has previously been reported also in patients with chronic heart failure (Xia 2009) and with coronary artery disease (Bailon et al. 2003).

A query to the PubMed NLM database conducted on the 5th of May 2016, using the MeSH (heart rate variability AND very high frequency) returned only two items beside the ones cited above (Özyılmaz et al. 2015; Pinhas et al. 2004). A more extended search has allowed us to find other few studies. The first reports of VHF HRV were from studies on cardiac transplant patients, investigating the possibility of autonomic re-innervation of the transplanted heart (Bernardi et al. 1989, 1990). Peaks in the VHF spectrum were considered a consequence of autonomic denervation. Mateo et al. (2001) have reported that the VHF band power spectral density (0.4–1.0 Hz) is useful in differentiating between ischemic and non-ischemic cardiac patients. The VHF and very low frequency (VLF) band indices also are relevant in patients with coronary ischemic disease during exercise (Bailon et al. 2003). Studies on reinnervation after cardiac transplantation demonstrate the VHF peaks in the spectra of HRV and blood pressure variability, but not in the spectra of a continuously recorded respiratory signal (Toledo et al. 2003). The presence of VHF activity can serve as a diagnostic test for vagal denervation. In patients with decreased left ventricular systolic function, higher values of a VHF HRV index have been seen, compared to healthy subjects (Xia 2009). In addition, pediatric patients who undergo transcatheter closure of atrial septal defects show an increase in VHF band (0.35–0.50 Hz) values one day after the closure. The VHF returns to the control level within 6 months of the intervention (Özyılmaz et al. 2015).

Although the number of studies on the VHF band is scant, consistent suggestions appear that this range of HRV is linked to physiological phenomena and may be useful as a clinical tool. The objective of the present study was to explore the VHF range of spectral frequencies of HRV using traditional as well as more recently developed non-linear and non-stationary spectral methods in a group of healthy subjects. These VHF were then compared with those in a group of SCA2 patients with a positive diagnosis of cardiovascular autonomic neuropathy to assess a potential clinical significance of the VHF evaluation.

2 Methods

2.1 Participants

The Ethics Committee of the Institute of Neurology and Neurosurgery of Havana in Cuba approved this study and deemed it as being conducted according to the standards of the Declaration of Helsinki. All participants gave written informed consent to participate in the study. Forty eight patients with SCA2, with a diagnosis of cardiovascular autonomic neuropathy (CAN) and no limitations for ambulatory treatment were included in the study. Eighty healthy non-smoker participants, without a history of cardiorespiratory or neurological disorder, diabetes mellitus, use of medications with known autonomic nervous system effects, and a normal 12-lead clinical electrocardiogram (ECG), were enrolled as a control group. Demographic and clinical features of all participants are shown in Table 1.

2.2 Clinical Sessions

All participants were studied from 08:00 a.m. to 12:00 p.m. They were instructed to abstain from physical efforts during the day before the study, to avoid drinking beverages containing caffeine, to sleep for at least 7 h the night before, to have their usual breakfast, and to drink a glass of fruit juice at least 1 h before the study. The SCA2 patients who smoked were asked to abstain from smoking after 07:00 p.m., the night before the study. All participants had to rest for 30 min sitting in a chair in a semi-reclining position, while the ECG electrodes were placed, during the equipment setting and calibration, and during recording quality testing. The temperature in the laboratory was maintained about 25 °C.

Autonomic tests used for the diagnosis of CAN were the cardiovascular reflex tests recommended by the Toronto Consensus Panel on Diabetic Neuropathy (Spallone et al. 2011), accompanied by the criteria added by others studying CAN, previously described (Cygankiewicz and Zareba 2013; Vinik et al. 2004, 2013; Vinik and Ziegler 2007; Boulton et al. 2005; Ewing et al. 1985). The cardiovascular reflex tests and normality limits considered for them were the following: 1/ *Standing test*: patients were asked to adopt the active standing posture for 10 min, after 15 min lying supine; the duration of the15th and the 30th inter interval beats after standing were measured to obtain the 30:15 ratio index. A value ≤ 1.0 was considered abnormal. Arterial blood pressure was measured, using auscultatory sphyngomanometry during the standing test (one minute prior to standing, 30 s after standing, and then every minute until the test-end); a fall >30 mmHg during the standing position was considered abnormal; 2/*Valsalva maneuver*: forceful expiration through a manometer, maintaining 40 mmHg for 15 s while sitting; the ratio of longest to shortest R-R during the test, known as the Valsalva index, was considered abnormal for a

Table 1 Demographic and clinical features of the investigated groups

	Control group ($n = 80$)	SCA-2 Patients ($n = 48$)
Age (years)	33.8 ± 9.7 [19–59]	37.4 ± 9.1 [25–61]
Gender (M/F)	40/40	24/24
Height (cm)	166.1 ± 9.0	167.6 ± 8.0
Weight (kg)	66.7 ± 10.4	64.2 ± 8.9
BMI (kg/m^2)	21.3 ± 0.03	19.9 ± 3.0
SBP (mmHg)	114.4 ± 14.0	136.8 ± 14.0**
DBP (mmHg)	71.3 ± 8.0	79.1 ± 9.0*
Years of disease	–	12.6 ± 5.0
ICARS	–	47.1 ± 5.0 [31–81]
CAG repeat length	–	36.8 ± 4.0 [35–46]
Age at disease onset	–	27.6 ± 8.0 [15–52]

Values are means ±SD
SBP Systolic blood pressure, *DBP* Diastolic blood pressure, *BMI* Body mass index, *[ ]* min-max, *p < 0.05; **p < 0.01 a *t*-test comparison between patient *vs.* control group, *ICARS* International Cooperative Ataxia Rating Scale, *CAG* Cytosine, adenine and guanine nucleotides triplets

value ≤ 1.10; 3/*Deep breathing*: participants breathed in and out while sitting, with a frequency of six breaths *per* minute paced by a metronome; a difference in heart rate < 10 bpm was considered abnormal. The lowest value for the expiration-to-inspiration ratio is dependent on age (Vinik et al. 2004). We also considered the signs and symptoms of the subjects during testing sessions. Time intervals between the tests were at least of 4-min duration. The methods, similar to those outlined above, have been previously employed in some of our other studies (Estévez-Báez et al. 2016a, b; Montes-Brown et al. 2010, 2012).

The stages of CAN were defined as early, definite, severe, and symptomatic following the established criteria (Spallone et al. 2011; Vinik and Erbas 2013). The abnormal result of one cardiovascular autonomic reflex test served as evidence of early CAN; at least two abnormal test results were required for a definite diagnosis of CAN; the presence of orthostatic hypotension in addition to any abnormal heart rate test identified severe CAN, and when cardiovascular symptoms accompanied the abnormal results of tests, the subject was in the symptomatic stage. We discarded symptomatic patients from participating in the study.

2.3   Recordings

2.3.1   ECG Recordings

ECG was recorded with commercial amplifiers (patient monitor Hewlett Packard 78354A; Palo Alto CA) and digitized with a 12-bit analog digital converter board (USB-6008 DAQ, National Instruments; Austin TX). A/D conversion was carried out with a sampling rate of 1 kHz. To control the process of digitization and storing of the ECG signal in the hard disk of PC, specific software was written into LabView 10.0 (National Instruments) by a staff member (J. J-G.). Filters were set for a band spectrum of 0.5–45 Hz. ECG signal was obtained from disposable electrodes placed on the chest in positions CM2 and V5.

ECG recordings were imported offline to a software tool developed in Delphi Embarcadero XP by members of our staff (M. E-B. and A. M-G., MultiTools version 3.1.2, 2009–2016) for visual inspection and detection of the fiduciary 'R' peaks. The accurate 'R' peak automatic detection was visually checked and properly corrected, if needed. Persons with isolated ventricular ectopic beats or supraventricular events were not included in this study.

Creating R-R Cardio-Interval (R-Ri) Time Series The original short ECG recordings were resampled by interpolation at 4 Hz with a cubic spline method included in the Matlab function 'interp1.m'. The resampling was necessary to transform the original R-Ri time series into a properly temporal series. This transformation resulted in an R-Ri sequence dataset for every subject, containing 1200 items with a sampling period of 0.25 s used for further processing.

2.3.2   Heart Rate Variability Calculations

Time Domain HRV Indices The mean values of the R-R interval series (MRR), the standard deviation of normal-to-normal R-R intervals (SDNN), and the root mean square of successive differences of R-R intervals (RMSSD), were calculated as recommended (Kuusela 2013; Task Force of ESC and NASPE 1996).

Intrinsic Heart Rate Estimation The intrinsic heart rate index was estimated in every subject with a classic method (José and Collison 1970) using the regression equation:

$$IHR = 118.1 - (0.57 \times age) \qquad (1)$$

2.3.3   Frequency Domain HRV Indices

Three spectral methods were used to avoid a bias. These methods will be conventionally referred in this study as Spa-Welch, Spa-EMD, and

Spa-Hilbert. The main steps of the spectral methods are summarized in Fig. 1.

Spa-Welch Spectral Method R-Ri resampled time series, subject to spline cubic interpolation with time duration of 300 s and with 1200 items, was submitted to two preprocessing actions. Firstly, linear detrending that consisted of a computation of a least-square fit of a straight line sequence to the data and subtracting the resulting function from the R-Ri series, using the Matlab function 'detrend.m'. This function included a demeaning of the R-Ri series, which contributed to a reduction of the DC component of R-Ri tachogram spectra. Secondly, high-pass filtering that consisted of the application of the order 6 Butterworth digital filter with a cut-off frequency at 0.02 Hz, using a zero-phase-shifting procedure with the Matlab function 'filfilt.m'. Spectral analysis of the preprocessed R-Ri series was carried out using a method of the smoothed periodogram of the Matlab function 'pwelch.m', having 2000 samples for the FFT process with the Hamming window. Segments of 200 samples with overlaps of 50% were used. These input parameters enabled the estimation of a power spectral density (PSD), averaging 11 individual spectra, with a spectral resolution of 0.002 Hz and 1001 spectral discrete frequencies. Using this spectral method it was possible to obtain nine discrete spectral frequencies for the very low frequency band of heart rate variability (VLF) from 0.02 to 0.04 Hz, 55 spectral frequencies for the low frequency band (LF) from 0.04 to 0.15 Hz, 126 spectral frequencies for the high frequency band (HF) from 0.15 to 0.40 Hz, and 250 spectral frequencies for the very high frequency band (VHF) from 0.4 to 0.9 Hz. We have previously employed the Welch method, albeit with a higher resampling frequency (Estévez-Báez et al. 2016a, b; Machado et al. 2014; Machado-Ferrer et al. 2013; Strobel et al. 1999).

Spa-EMD Spectral Method Resampled R-Ri series were submitted to a nonlinear, non-stationary, and data-driven decomposition into monotonic intrinsic mode functions (IMFs), using the (EMD) algorithm (Huang et al. 1998). The algorithm is known as complete ensemble empirical mode decomposition with adaptive noise (CEEMDAN) (Colominas et al. 2014), with a recent modification of Torres et al. (2011). The implementation of this algorithm in the Matlab code in the present study took place with permission from http: www.bioingenieria.edu.ar/grupos/ldnlys.

To consider the IMF as corresponding to a valid methodological frequency, following the concepts of harmonic theory, only were IMFs

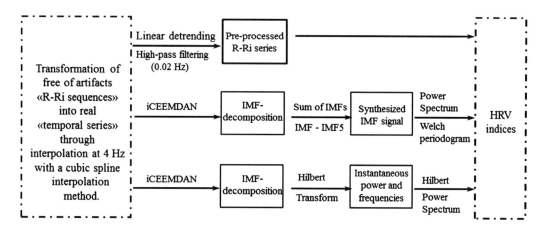

Fig. 1 Flow chart of the spectral methods applied in the study. The upper sequence of events indicates the steps carried out in the Spa-Welch spectral method. The middle sequence depicts the steps of the empirical mode decomposition (Spa-EMD) spectral method, and the bottom sequence represents the steps of the Spa-Hilbert spectral method. *HRV* Heart rate variability, *IMF* Intrinsic mode function

accepted which showed at least ten complete cycles for total series duration of 300 s. This recommendation is akin to that for the valid frequencies that can be considered for HRV spectral analysis (Task Force of ESC and NASPE 1996). The added values of extracted IMFs created a signal that contained all valid frequencies from the original R-Ri sequence. This signal was submitted directly to spectral analysis using the Welch smoothed spectral periodogram.

Spa-Hilbert Spectral Method The R-Ri series resampled by interpolation was decomposed in intrinsic mode functions, and each of the extracted IMF was submitted to the Hilbert transformation. This method can be applied to IMFs, because the signals are monotonic (mono-component) (Huang et al. 1998). For each Hilbert transformed IMF, the instantaneous frequencies and their corresponding values of energy, expressed as the power spectral density, were calculated.

The concept of an analytic signal is essential to explain the notion of instantaneous indices obtained from IMFs. This is a complex signal with one-sided spectrum that preserves all the information contained in the original signal (Colominas et al. 2014). The application of the Hilbert transform is a way to obtain the analytic signal, hitherto referred to as W(t):

$$W(t) = X(t) + Y(t) \qquad (2)$$

where X(t) is the input the time series i.e., the IMF in this case, and Y(t) the Hilbert transform of X(t).

Now, several instantaneous indices can be obtained using the following expressions:

$$P(t) = \left[X(t)^2 + Y(t)^2\right]; \quad \theta(t)$$
$$= \arctan\left(\frac{Y(t)}{X(t)}\right); \quad \omega(t) = \frac{d\theta(t)}{d(t)} \qquad (3)$$

where P(t) is the instantaneous energy power, $\theta(t)$ is the instantaneous phase, and $\omega(t)$ is the instantaneous frequency.

The instantaneous values of power spectral density and spectral frequency of each extracted IMF were used for the ulterior calculations of spectral indices with this method. While the Spa-Welch and Spa-EMD methods are limited to the analysis of spectral frequencies that are multiples of the spectral resolution of 0.002 Hz, the Spa-Hilbert method is not affected by this limitation, because the number of instantaneous spectral frequencies amounts in this case is 1200 discrete frequencies and their corresponding PSD values for each extracted IMF. These values were included in a two-column matrix. As for each of the five valid IMFs 1200 values were obtained, the Hilbert power spectral matrix contained 6000 instantaneous frequencies and PSD values.

The Hilbert marginal spectra were calculated to graphically represent the smoothed Hilbert power spectra. For these calculations, a function in the Matlab code was developed by one of the authors (M. E-B.). The function enables the selection of a desired frequency resolution for the spectrum that can be useful for visual and quantitative comparisons with the power spectra obtained through methods with fixed values of discrete spectral frequencies, such as different FFT-based procedures, including the Spa-Welch and Spa-EMD methods.

2.3.4 Spectral Indices

The estimations of the power spectral density expressed in the absolute values, relative PSD expressed as normalized values (%), and the mean frequency values of PSD detected in the different frequency bands were calculated from the power spectra obtained with the Spa-Welch and Spa-EMD methods, and from the power spectral matrixes obtained with the Spa-Hilbert method. The three expressions used for the calculations were as follows:

$$E(abs) = \sum_{i=min}^{max} DF_i \qquad (4)$$

where 'min' is the lower limit value of the spectral range of a band, 'max' identifies the upper limit, and DF_i are the discrete spectral frequencies. The acronyms for these indices were named as P-VHL for the very low frequency band; P-LF for the low frequency band; P-HF for the high frequency band; P-VHF for the very high frequency band, and P-Total for the whole power spectral density in the frequency range of 0.02–0.90 Hz.

$$E(rel) = \frac{E(abs)_{band}}{E(Total)} {}^{*} 100 \qquad (5)$$

where $E(abs)_{band}$ is the absolute energy of a given band and $E(total)$ is the energy sum of all spectral bands. The acronyms for these indices were the same as those for the corresponding used before, but were preceded the prefix 'nu-', meaning normalized units. Thus, we included the nu-VLF, nu-LF, nu-HF, and nu-VHF indices.

$$mean(f) = \sum_{k=min}^{max} \frac{f_k {}^{*} Eabs_k}{E(abs)_{band}} \qquad (6)$$

where min and max have the same meaning as those in Expression 3, and f_k is the value of a corresponding discrete spectral frequency. The acronyms for each band were here preceded by the 'mf-'prefix. We calculated the mf-VLF, mf-LF, mf-HF, and mf-VHF indices.

The ratio of the absolute values in the LF band and HF bands also was calculated and the acronym for this index was LF/HF. The mean instantaneous frequency values, calculated from the Hilbert transformations of the extracted IMFs, were calculated for each R-Ri series of all the subjects in both groups. Grand averages of the spectra obtained with each of the three methods were calculated for both groups for the 0.02–0.90 Hz and 0.40–0.90 Hz ranges. A mean value ±SE was calculated for each discrete spectral frequency in the studied range and was superimposed to the grand mean PSD for a visual analysis as previously described (Estévez-Báez et al. 2016a, b; Machado-Ferrer et al. 2013; Marple 1999). All digital signals were processed using the standard custom-tailored programs developed by our staff, based on Matlab software (R2015a, 8.5.0.197613; MathWorks Inc., MA).

2.4 Statistical Analysis

Data were presented as means ±SD, unless otherwise indicated. Data distribution was assessed with the Shapiro-Wilk test. Corrections to achieve normality consisted of natural or common log transformations. A *t*-test for independent samples was used to compare demographic and clinical indices between the groups. Two independent samples of normally distributed values were compared with one-way ANOVA and Fisher's LSD *post hoc* test. The ANOVA also was applied for differences between the healthy subjects and SCA2 patients as well as for the HRV indices in the time domain. A univariate ANOVA design was used to assess differences in the spectral indices, with a two-level inter-group factor (Group), a three-level within-subject repeated measures factor (Method), and a Group-Method interaction. The statistical power for each test was considered valid only for values over 0.7, and the *post-hoc* comparisons were conducted with Fisher's LSD test for factor Group and the Newman-Keuls test for factor Method. A p-value <0.05 was accepted as significant. The relationship between the calculated indices of HRV was assessed with Pearson's product-moment correlation index. A commercial STATISTICA v8.0 package (StatSoft Inc., Tulsa, OK) was used for all analyses.

3 Results

3.1 Demographic and Clinical Indices

Factors age and gender showed no differences between the groups (Table 1). No significant differences were found for height, weight, and body mass index. Systolic and diastolic blood pressure was significantly higher in the group of patients.

3.2 Autonomic Tests

Nine out of the 48 patients were diagnosed with severe CAN (18.8%), 23 patients presented a definite CAN (47.9%), and 16 patients (33.3%) presented an early CAN. The control healthy subjects did not present any kind of abnormal responses to the autonomic testing. A graphical representation of these results for SCA2 patients is presented in Fig. 2.

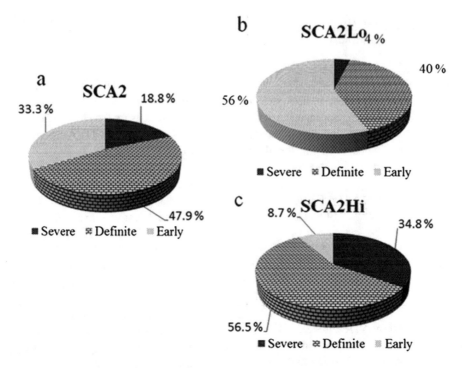

Fig. 2 (**a**) Graphic representation of the severity of the cardiovascular autonomic neuropathy in the whole group of patients, and stratified into two subgroups: (**b**) those with nu-VHF values lower (SCA2Lo) and (**c**) nu-VHF values higher (SCA2Hi) than the 99th percentile of the values observed in healthy subjects; *nu* stands for normalized units. *SCA2* spinocerebellar ataxia type 2

3.3   Time Domain HRV Indices

The mean heart R-R interval (MRR) and the global and beat-to-beat variability indices SDNN and RMSSD were significantly reduced in the patient group (Fig. 3a–c). The calculated values of the intrinsic heart frequency (IHF) did not show significant differences between the controls (94.4 $\pm$ 7.1 beats/min) and the SCA2 patients (95.0 $\pm$ 5.1 beats/min) (F(1,126) = 0.22, p = 0.65).

3.4   Frequency Domain HRV Indices in the Conventional Frequency Ranges

The empirical mode decomposition in intrinsic mode functions (IMFs) obtained from the R-Ri tachograms of a control subject and a SCA2 patient, which illustrate the most typical patterns found in all subjects, are shown in Fig. 4. The tachogram in the healthy subject (Panel A) shows higher values and a greater variability than those in the SCA2 patient (Panel B). There also is an increase in the high frequencies particularly in the IMF-1 and IMF-2 components, and a reduction in the amplitude of cyclical oscillations for the IMFs from the IMF-1 to the IMF-5 in the SCA2 patient (Panel B). Eight IMFs were extracted in the healthy subject, while nine IMFs were extracted in the patient. Nevertheless, only were the first five IMFs valid in both subjects for ulterior analysis, because after the fifth IMF the extracted components showed fewer than ten complete cycles for the whole duration of 300 s.

The spectra corresponding to the subjects in Fig. 2, evaluated with the Spa-EMD method and the Spa-Hilbert method, are shown in Fig. 5. The values of relative energy contents of spectral bands, expressed as normalized units (%) in bar plots, are shown for each spectrum. The spectra obtained with both methods show similarities, and almost the same distribution of energy content. A reduction

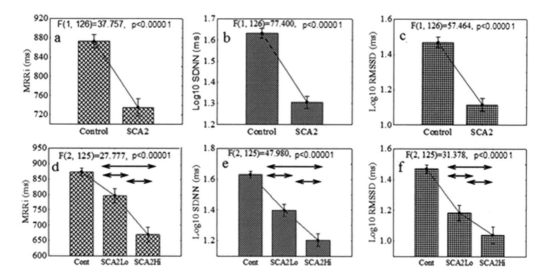

Fig. 3 Comparison of heart rate variability (HVR) in time domain, the mean heart R-R interval (MRR), the standard deviation of normal-to-normal R-R intervals (SDNN), and the root mean square of successive differences of R-R intervals (RMSSD) in healthy controls and spinocerebellar ataxia type 2 (SCA2) patients. Diagrams a, b, and c show the comparison between the groups as a whole, while diagrams d, e, and f show the comparisons between the control healthy group and the SCA2 patients stratified into those with nu-VHF values higher (SCA2Hi) or lower (SCA2Lo) than the 99th percentile of the values observed in the healthy subjects. The values corresponding to ANOVA Fisher's F statistic are included in each diagram. Double arrow lines represent significance at p < 0.0001. Vertical bars denote ±SE

of the nu-HF band and an increment in the nu-VHF in the SCA2 patient are worthy of note.

The results of a comparative analysis of the control and SCA2 patient groups, using the three spectral methods, are shown in Table 2. The univariate ANOVA shows that the estimated absolute values of PSD for the classical HRV bands VLF, LF, and HF, including the total PSD for the three bands, were significantly lower in the SCA2 patients compared with the control group. The relative energy content decreased in the HF band while the LF/HF ratio increased in the patients. The statistical power was over 0.9 for all the mentioned HRV indices and the significance of differences between the groups included in Table 2 was assessed with the *post hoc* Fisher LSD test.

3.5 Periodogram Averaging

The grand average of the smoothed Hilbert marginal power spectra of the healthy subjects obtained with the Spa-Hilbert method is shown in Fig. 6a, while the corresponding grand average for the whole group of SCA2 patients is in Fig. 6c. The PSD magnitude evidently differed between both groups, being larger for the control healthy group. Also, PSD in the HF range was significantly reduced in the patients compared with controls. Although the PSD magnitude was very low in the range over 0.40 Hz compared with that of 0.02–0.40 Hz, a discrete increment of the VHF components could be seen from 0.4 to approximately 0.48 Hz.

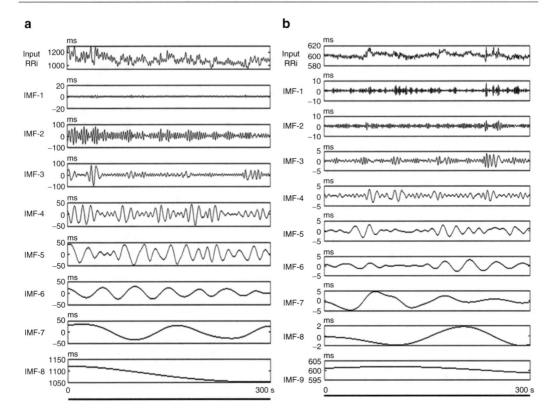

Fig. 4 Examples of empirical mode decomposition (EMD) applied to R-Ri series of 5-min length using the improved CEEMDAN algorithm in a healthy subject (Panel A) and in a spinocerebellar ataxia type 2 (SCA2) patient diagnosed with cardiac autonomic neuropathy (Panel B)

3.6 Frequency Domain HRV Indices in the Very High Frequency Range

The most outstanding difference between the healthy subjects and SCA2 patients was a significant increment in power spectral density in the latter group, expressed in normalized units, in the VHF range (index nu-VHF) (Table 2). The mean frequency index mf-VHF showed significantly higher values in the patients.

Only were the first five IMFs, extracted with the improved CEEMDAN algorithm for the EMD process, found valid in the frequency analysis in both groups. The mean values of the instantaneous frequencies, corresponding to these IMFs,

and the ranges for the mean interval $\pm$ 2SD are shown in Table 3. Highly significant differences, with larger values for the SCA2 patients, were found for the mean instantaneous frequencies of the first three intrinsic mode functions. The mean instantaneous frequencies of the fourth and fifth IMF did not show significant differences between the groups. There was some correspondence of the IMFs with the classical HRV bands (Table 3). The first IMF included only the frequencies of the VHF band in the limits explored in this study. The IMF-2 included frequencies corresponding to the VHF band and also to the HF band. The IMF-3 included only frequencies of the HF band. The IMF-4 showed frequencies mainly of the LF

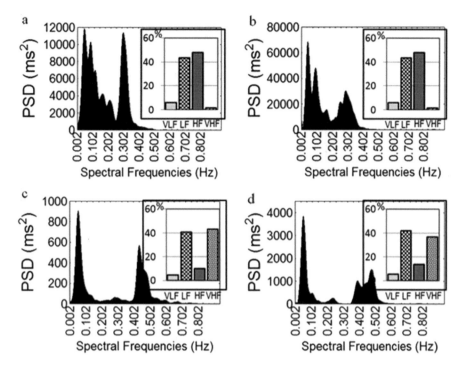

Fig. 5 Power spectra and relative energy content, shown as bar diagrams of HRV bands, corresponding to the healthy subjects (upper diagrams **a** and **b**) and spinocerebellar ataxia type 2 (SCA2) patients (lower diagrams **c** and **d**) shown in Fig. 2. Spectra in diagrams a and c were obtained with the spectral empirical mode decomposition method (Spa-EMD), while those in diagrams b and d are the Hilbert smoothed power marginal spectra obtained with the Spa-Hilbert method. *PSD* Power spectral density

band, IMF-5 showed frequencies corresponding to the LF and VLF bands of HRV.

Since the most outstanding difference, found between the control and patient groups, consisted of the increment in the nu-VHF index, the value corresponding to the 99th percentile (P_{99}) for this index in the control group was calculated. The patients with values higher or lower than that value were identified. A subgroup 23 of SCA2 patients (47.9%) had higher values and was named the SCA2Hi subgroup. The remaining 25 patients (52.1%) with lower values, were named the SCA2Lo subgroup. The severity of cardiovascular autonomic neuropathy was more evident in the SCA2Hi patients than in the SCA2Lo patients, as seen in Fig. 2b, c. Differences between these two subgroups of patients were also shown for the time domain indices of HRV (Fig. 3).

The averaged periodograms also showed evident differences for both subgroups of patients.

The grand averages of the Hilbert smoothed power marginal spectra, depicted in Fig. 5b, d, show that the power spectral density values were greater in the SCA2Lo subgroup than in the SCA2Hi subgroup. A comparisons between the three spectral indices P-VHF, nu-HF, and mf-VHF, considering both the whole groups (diagrams a, b, and c) and the subgroups of patients (diagrams d, e, and f) are presented in Fig. 7. A significant increment in the nu-VHF index was only detected between the controls and either the SCA2Lo or SCA2Hi patients, but not between these two patient subgroups. For the mf-VHF index, significantly higher values were found for the SCA2Hi subgroup compared with the SCA2Lo subgroup, and between both of them against the controls.

As the IMF-2 extracted component contains the frequencies of the HF band, the index mf-HF was included in comparative analysis between the whole groups (Table 2). The mean frequency

Table 2 Comparative analysis of frequency domain HRV indices in the healthy subjects and spinocerebellar ataxia type 2 (SCA2) patients diagnosed with cardiac autonomic neuropathy (CAN)

Indices	Spa-Welch method		Spa-EMD method		Spa-Hilbert method	
	Controls ($n = 80$)	SCA2 ($n = 48$)	Controls ($n = 80$)	SCA2 ($n = 48$)	Controls ($n = 80$)	SCA2 ($n = 48$)
Log_e P_VLF (ms^2)	10.29 ± 1.18	9.35 ± 1.05***	10.21 ± 1.03	8.47 ± 1.30***	11.67 ± 1.19	9.68 ± 1.59***
Log_e P_LF (ms^2)	12.45 ± 0.99	10.85 ± 1.32***	12.37 ± 1.01	10.73 1.41***	13.70 ± 1.01	12.10 ± 1.45***
Log_e P_HF (ms^2)	11.97 ± 1.13	9.78 ± 1.53***	11.99 ± 1.13	9.78 ± 1.52***	13.15 ± 1.17	10.92 ± 1.58***
Log_e P_Total (ms^2)	13.18 ± 0.94	11.51 ± 1.21***	13.05 ± 0.98	11.38 ± 1.26***	14.38 ± 0.98	12.68 ± 1.30***
nu_VLF (%)	11.2 ± 5.20	13.29 ± 6.20	6.83 ± 4.20	6.80 ± 4.10	9.33 ± 7.10	7.44 ± 5.40
nu_LF (%)	51.3 ± 15.60	60.26 ± 15.00**	52.24 ± 16.70	55.93 ± 18.00**	54.24 ± 17.10	59.43 ± 18.10**
nu_HF (%)	34.6 ± 16.60	21.44 ± 13.70***	37.73 ± 16.90	24.24 ± 4.40***	34.44 ± 17.90	21.50 ± 14.10***
LF/HF ratio	2.27 ± 2.05	4.17 ± 3.63***	2.11 ± 1.94	3.82 ± 3.56***	2.72 ± 2.89	5.15 ± 5.47***
mf_HF (Hz)	0.25 ± 0.03	0.28 ± 0.03**	0.25 ± 0.03	0.29 ± 0.03**	0.25 ± 0.04	0.29 ± 0.03*
Log_e P_VHF (ms^2)	9.39 ± 0.98	8.92 ± 0.97***	9.39 ± 0.96	8.92 ± 0.97**	10.05 ± 1.07	9.95 ± 1.07
nu_VHF (%)	2.80 ± 1.09	10.91 ± 9.60***	3.15 ± 2.10	12.98 ± 12.70***	1.99 ± 1.70	11.63 ± 12.60***
mf_VHF (Hz)	0.49 ± 0.03	0.50 ± 0.05***	0.49 ± 0.03	0.54 ± 0.05***	0.47 ± 0.03	0.52 ± 0.05***

Values are means ±SD

Spa-EMD Empirical mode decomposition (EMD) spectral method. Acronyms of different indices are described in the Method section

Fisher's LSD post hoc tests for comparisons between groups: *p < 0.05; **p < 0.01; ***p < 0.001

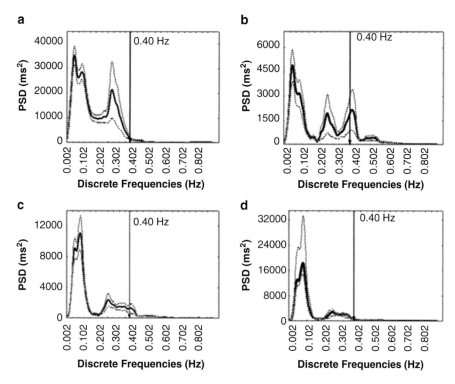

Fig. 6 Grand averages of the Hilbert smoothed marginal power spectra (continuous bold lines) for: (**a**) 80 control healthy subjects, (**b**) 23 spinocerebellar ataxia type 2 (SCA2) patients with nu-VHF values higher than the 99the percentile (P99) of the values observed the control subjects (SCA2Hi); (**c**) grand average for the whole group of SCA2 patients, and (**d**) 25 SCA2 patients with nu-VHF values lower than the 99the percentile (P99) of the values observed in the control subjects (SCA2Lo). Dotted lines all diagrams represent the distribution of the standard error of the means for every discrete spectral frequency. The vertical arrows point to the frequency limit 0.40 Hz of the traditional HF band

Table 3 Comparison of mean instantaneous frequencies calculated with the Hilbert transform for the first five intrinsic mode functions (IMFs) extracted from the R-Ri series between the control healthy and spinocerebellar ataxia type 2 (SCA2) patients

Indices	Controls (n = 80)	SCA2 Patients (n = 48)	HRV band	F(1,126)	p
mif-IMF-1	0.58 ± 0.05 [0.49–0.68]	0.64 ± 0.07 [0.51–0.77]	VHF	27.85	0.001
mif-IMF-2	0.36 ± 0.06 [0.25–0.47]	0.46 ± 0.08 [0.30–0.62]	HF & VHF	65.63	0.001
mif-IMF-3	0.23 ± 0.03 [0.17–0.29]	0.28 ± 0.04 [0.12–0.36]	HF	57.86	0.001
mif-IMF-4	0.11 ± 0.01 [0.09–0.13]	0.11 ± 0.01 [0.03–0.13]	LF	0.55	0.460
mif-IMF-5	0.05 ± 0.01 [0.04–0.07]	0.06 ± 0.11 [0.05–0.07]	LF & VLF	0.50	0.480

Values are means ± SD

mif Mean instantaneous frequency, *IMF* Intrinsic mode function, *[ ]* range for mean values ±2SD, *HRV band* corresponding HRV bands for the frequency ranges, *F* Fisher's ANOVA statistic
p associated probability for the F values

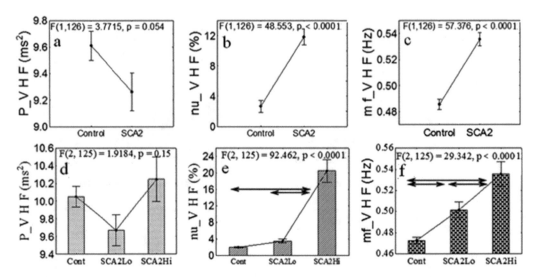

Fig. 7 Upper row diagrams are: (**a**) P-VHF, (**b**) *nu-VHF* normalized units frequency index, and (**c**) *mf-VHF* mean frequency index. The corresponding F-statistic values for univariate ANOVA are depicted in the upper border of each diagram. Highly significant differences were obtained for the nu-VHF and the mf VHF indices (Scheffe's *post-hoc* test; p < 0.00001). Lower row diagrams (**d**, **e**, and **f**) correspond to the one-way ANOVA results between the healthy controls and the spinocerebellar ataxia type 2 (SCA2) patients stratified into the SCA2Lo and SCA2Hi subgroups. *VFH* very high frequency. F-statistic values are here also depicted in the upper border of each diagram. Double arrow lines represent significant values for p < 0.0001 for Fisher's *post hoc* LSD tests. Data are means ± SE

values for the lower frequency bands of LF and VLF are not shown since those bands were not directly related to the results observed in the VHF band. Significantly greater mf-HF values were detected for the whole SCA2 patient group, and also for both SCA2Lo and SCA2Hi subgroups of patients against the control group, but not between the patient subgroups (Fig. 8e, f). The mf-VHF index also yielded greater values between either of the patient subgroups and the control group, and between the two patient subgroups, with greater values for the SCA2Hi patients (Fig. 8g, h). The mean instantaneous frequency value for the IMF-1 was significantly greater in either patient subgroup than that in the controls, with an inappreciable difference between the two subgroups (Fig. 8a, b). Likewise,

the mean instantaneous frequency value for the IMF-2 also was significantly greater in either patient subgroup than that in the controls, but the difference between the two patient subgroups was here appreciable, with the greater values in the SCA2Hi subgroup (Fig. 8c, d).

A correlation analysis showed a low but significant relationship between the intrinsic heart rate index and P-VHF in the controls and SCA2 patients (Table 4). Heart rate was related to almost all the HRV spectral indices of PSD. The mean instantaneous frequencies for the components IMF-1 and IMF-2 also showed significant correlation coefficient values, which were greater in the patients. Correlation coefficients of the mean frequency values of mf-HF and mf-VHF showed a similar profile.

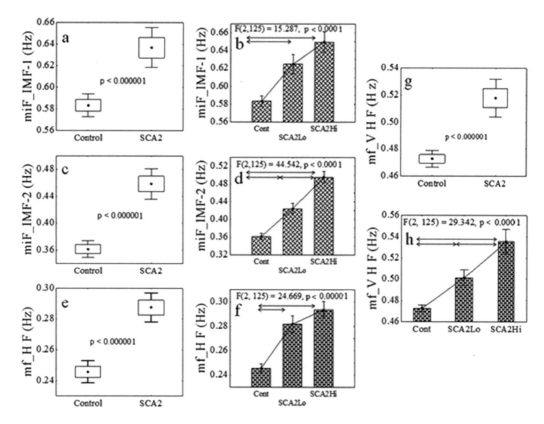

Fig. 8 Comparisons between frequency spectral indices, calculated with the Spa-Hilbert spectral method, for the whole groups of controls and spinocerebellar ataxia type 2 (SCA2) patients, and between the control subjects and either subgroup of patients (SCA2Lo and SCA2Hi). Double arrow lines indicate significant values at $p < 0.0001$ for Fisher's *post hoc* LSD tests. Data area means $\pm$ SE. (**a–h**) identify the diagrams corresponding to specific HRV indices

Table 4 Pearson's correlation coefficients (r) calculated between frequency spectral HRV indices and absolute, relative HRV indices of the estimations of power spectral density in the control and spinocerebellar ataxia type 2 patient groups. Results are presented as r (controls)/r (patients)

Indices	IHR	HR	mif-1	mif-2	mf-VHF	mf-HF
P–VLF	0.23/0.05	−0.51/−**0.64**	−**0.36**/−**0.39**	−0.01/−**0.34**	−0.23/−**0.48**	−0.23/−0.25
P–LF	**0.54**/0.30	−0.29/−**0.71**	−0.29/−**0.55**	0.11/−**0.40**	−0.09/−**0.55**	−0.35/0.36
P–HF	**0.45**/0.21	−**0.52**/−**0.80**	−**0.48**/−**0.72**	−**0.54**/−**0.75**	−**0.37**/−**0.79**	−0.00/−0.05
P–VHF	**0.35**/**0.39**	−0.08/−0.15	−0.02/−**0.50**	0.18/−0.03	−0.04/−0.10	0.18/−0.10
P–Total	**0.53**/0.28	−**0.44**/−**0.76**	−**0.44**/−**0.66**	−0.15/−**0.51**	−0.22/−**0.66**	−**0.23**/−0.27
nu–VLF	−**0.31**/−0.28	−0.14/0.06	−0.05/0.18	0.15/0.20	−0.03/−0.04	−0.05/−0.04
nu–LF	0.13/0.24	**0.35**/−0.10	**0.30**/0.14	**0.70**/0.28	**0.36**/0.15	−**0.39**/−**0.46**
nu–HF	0.00/−0.04	−**0.30**/−**0.42**	−0.29/−**0.46**	−**0.75**/−**0.73**	−**0.35**/−**0.60**	**0.35**/**0.32**
nu–VHF	−0.07/−0.18	**0.22**/**0.58**	0.21/0.24	**0.23**/**0.33**	0.14/**0.39**	**0.44**/**0.32**
LF/HF	−0.03/0.08	**0.28**/0.14	0.17/0.15	**0.73**/**0.46**	**0.36**/**0.28**	−**0.33**/−**0.37**

IHR Intrinsic heart rate index, *HR* Hear rate; *mif-1* Mean instantaneous frequency of the IMF-1 component, *mif-2* Mean instantaneous frequency of the IMF-2 component, *mf-VHF* Mean frequency in very high frequency band, *mf-HF* mean frequency in high frequency band. For other acronyms see Methods. Values in bold depict correlation coefficients of statistical significance $p < 0.05$

The possible effect of the resampling frequency used in this study (4 Hz) was assessed in comparison with previous calculations using a resampling frequency of 2 Hz for the interpolation of the R-Ri series. The spectral methods used for this assessment were the more classic Spa-Welch and the Spa-EMD methods. There were no significant differences between the 2 Hz and 4 Hz resampling frequencies using both methods in the absolute values of power spectral density, LF/HF ratio, and in the indices of mean spectral frequencies such as mf-VF or mf-VHF. Only did the Spa-EMD method show a significant difference in the relative (%) energy content of HRV bands, where the values obtained with resampling frequency of 2 Hz were significantly lower than those with 4 Hz. In the control group, nu-VHF index was 2.4 ± 1.6% *vs.* 3.2 ± 2.1, nu-HF was 29.6 ± 15.3% *vs.* 37.7 ± 16.9, and nu-LF was 43.4 ± 14.7% *vs.* 52.2 ± 16.7, with an exception for significant increase for the VLF band (24.3 ± 11.1% *vs.* 6.8 ± 4.2%) for 2 Hz and 4 Hz resampling frequency, respectively. Similar differences were observed for the SCA2 patients, where nu-VHF was 8.8 ± 7.5% *vs.* 13.0 ± 12.7%, nu-HF band was 17.2 ± 12.4% *vs.* 24.2 ± 4.4%, LF band was 55.9 ± 18.0% *vs.* 43.9 ± 16.5%, and an increase for the VLF band from 29.6 ± 14.3% to 6.8 ± 4.1% for 2 Hz and 4 Hz resampling frequency, respectively. The three spectral methods used in this study showed, generally, the same pattern of variation in results, albeit the values differed, depending on a given methodological calculation procedure.

4   Discussion

In the present study we investigated the time and frequency domains of HRV, using three different analytical methods, in SCA2 patients in comparison with healthy subjects. The findings were that the patients a substantial reduction in the parasympathetic cardiac input and, to a lesser degree, a fall in sympathetic cardiovascular control activity, which were associated with a general reduction in the global autonomic influence on the cardiovascular system. These findings were in

line with those previously reported and discussed by Montes-Brown et al. (2010, 2012) and Piskorski et al. (2010) in SCA2 patients and also in presymptomatic carriers of SCA2. Nevertheless, it was worthwhile to confirm that the three methods employed in the present study yielded a similar pattern of results, albeit they could be somehow different depending on the procedure used. The general profiles of changes and the statistical significance between the groups were similar for the three methods. These results, obtained with non-conventional spectral approaches, show that the classical and hitherto extensively used approach is perfectly valid and there is no reason to any modification, at least for the spectral indices calculated for a frequency range of 0.02–0.40 Hz. Therefore, we dwell on the VHF results in further discussion.

4.1   Cardiac Autonomic Neuropathy

Since the cardinal finding of Ewing et al. (1985) showing that diabetic patients with a positive diagnosis of CAN have a double risk of death after the second year into the disease, a spate of cardiovascular tests have been extensively used and their role has not been lost with time (Lacigova et al. 2016; Wheelock et al. 2016; Balcioglu and Muderrisoglu 2015; Chyun et al. 2015; Zoppini et al. 2015; Pradhan et al. 2007). In fact, CAN has been detected in many other diseases and conditions, including several types of the spinocerebellar ataxias, which increases CAN's importance in the assessment of cardiac death risk (Mukaino et al. 2016; Rasheedy and Taha 2016; Xu et al. 2016; Alonso et al. 2015; Montes-Brown et al. 2010; Boulton et al. 2005). CAN produces an autonomic denervation of the cardiovascular system and particularly of the heart, which is responsible of lethal complications in patients suffering from different diseases. Its course may be asymptomatic and it presence is usually detected with the use of cardiovascular reflex tests. Thus, the search for tools helping diagnose CAN cannot be underestimated.

4.2 Dynamics of VHF Indices

The present study shows that a distinct spectral activity of very low magnitude could be detected in a range of 0.4–0.9 Hz in healthy human subjects. The three spectral methods used corroborated this finding, but the Spa-Hilbert method, not only detected the very fast oscillations of HRV, but also enabled a more accurately study of their frequency content. A proportion of power spectral density (PSD) amounted to 0.12–6.93% in the percentile range from $P_{2.5}$ to $P_{97.5}$, respectively, in the VHF range studied in healthy subjects. A high inter-individual variability was observed when the values were expressed as means $\pm$ SD (1.99 $\pm$ 1.70%) as shown in Table 2. The mean spectral frequency, calculated from expression 6, was 0.47 $\pm$ 0.03 Hz, while the mean instantaneous frequencies showed 0.58 $\pm$ 0.05 Hz for the first extracted intrinsic mode function (IMF-1) component of the empirical mode decomposition (Tables 2 and 3). Only have two previous studies reported the mean and instantaneous spectral frequencies quantitatively for the PSD of VHF components, using a traditional FFT power spectrum. A value of 11.37 $\pm$ 10.77% is reported for the VHF index defined as the very high frequency component of the power spectrum normalized to represent its relative value in proportion to the total power minus the very low frequency component (Xia 2009) in a group of 64 patients with a cardiac disease who were used as controls for patients with reduced left ventricular systolic function. A value of 6.2 $\pm$ 9.1% is reported for 15 healthy subjects, and 9.99 $\pm$ 6.47% for 30 healthy children in a VHF range of 0.35–0.50 Hz (Özyılmaz et al. 2015). The present results were significantly lower for 80 healthy subjects, but showed a similar high inter-individual scatter as indicated by the standard deviation of the mean values.

In the present study we found that the PSD, expressed in absolute values, was significantly lower in SCA2 patients when it was calculated with the Spa-Welch and Spa-EMD methods, but not so when compared with the Spa-Hilbert

calculation. These results were akin to those reported in children with a closure of atrial septal defect and in healthy children using a traditional approach: 9.99 $\pm$ 6.47% vs. 11.56 $\pm$ 6.95% (Özyılmaz et al. 2015). A notable finding of the present study was a highly significant difference observed in PSD, expressed in normalized values, between the controls and SCA2 patients. A significant increase in the normalized PSD values in patients was in consonance with similar increases reported in cardiac patients with or without signs of reduced left ventricular systolic function; 11.37 $\pm$ 10.77% vs. 19.17 $\pm$ 13.35%, respectively (Xia 2009) and between healthy children and children with a closure of septal defect a day after the surgical procedure; 9.99 $\pm$ 6.47% vs. 18.48 $\pm$ 11.01%, respectively (Özyılmaz et al. 2015).

Previous studies have failed to report quantitative differences in the measures of VHF spectral frequencies. The present study demonstrates significantly higher values of the mean VHF frequencies and mean instantaneous frequencies of the extracted IMF-1 and IMF-2 components in SCA2 patients (Table 3). A stratification of patients according to a higher or lower value of the nu-VHF index, with a cut-off value of the 99th percentile (P_{99}) present in the healthy group, enabled the distinguishing of a relationship between the values for this index and the severity of cardiovascular autonomic neuropathy (Fig. 2). Likewise, there were differences between the two subgroups of patients in the time domain of HRV (Fig. 2) and in the VHF indices (Fig. 8). The subdivision of SCA2 patients also was useful for differentiation between either patient subgroup and the group of control subjects using the method of averaging periodograms (Fig. 6), and for finding a different behavior of the mean frequency values and the instantaneous mean frequency values, calculated from the power spectra and from the Hilbert transforms. We consider these results as evidence of the potential pathophysiological relationship of these newly described VHF indices, which speaks against the possibility of considering the VHF components described in this study as simple artifacts.

We could not find differences in the calculated intrinsic heart rate index between the healthy subjects and patients. However, correlation analysis showed a significant relationship for both controls and patients between the intrinsic heart rate index and the absolute power spectral density of the VHF band (P-VHF) (Table 4), while there was no correlation between the P-VHF and the values of the HR index in either group.

The EMD method brought to light the complexity of the frequency content in the first three extracted IMFs, which deserves further investigation. The frequency decomposition using this approach seems promising. Beside the VHF band, the presence of three other main bands in short-time HRV assessments, i.e., VLF, LF, and HF, is reliably detected (Task Force of ESC and NASPE 1996). Of note, the EMD method is a data-driven process, not needing a selection of any pre-defined parameters to carry out the analysis, which contrasts with different standard FFT spectral approaches, the Lomb-Scargle periodogram, different time-frequency methods, or the wavelets (Machado et al. 2011). Table 3 shows that the results obtained from the IMF-4 in the control subjects include a frequency range of 0.09–0.13 Hz, which corresponds to the classical LF band, but also is considered by many authors as a mid-frequency subrange, having particular physiological characteristics that point to an independent meaning (Widyanti et al. 2013; Middleton et al. 2011; Zhang et al. 2011).

4.3 Methodological Issues Associated with the Investigation of the VHF Range

Original R-Ri values are not exactly temporal series. They are ordinal series of values that can be described as a non-uniformly sampled event sequence. This fact was ignored by some pioneers of the HRV investigation (Baevskii et al. 1971; Zhemaitite 1967; Parin et al. 1965a, b), who have introduced a raw sequential ordered R-Ri series to the analysis of frequency content using autocorrelation methods, and the then developed FFT-based methods for spectral analysis. Those

methodological conflicts have been disclosed in a paper of Nidekker (1981), who have proposed a way to circumvent the problem. A time-window with a duration selected by the user enabled the calculation of cardiac cycles, or parts thereof, included in a window in the ECG recording. That made it possible to obtain equally sampled values of heart frequency, expressed as cardiac cycles *per* unit of time, i.e., window's duration.

Berger et al. (1986) have presented a method for transforming the sequential R-R series into temporal ones, affirming that any arbitrary sampling rate of the recorded ECG can be chosen. A local window could then be defined, including the initial and the following sample point. Then, the number of R-R intervals, or fractions thereof, is counted and the heart rate for each window is calculated using a predefined expression. Task Force of ESC and NASPE (1996) have accepted Berger et al.'s (1986) procedure, but recommended the use of the interpolation method for resampling the original sequential R-R series, with the provision that the resampling frequency should be sufficiently high. In practice, the sampling frequency employed has the values fluctuating between 4 and 10 Hz. DeBoer et al. (1984) have shown that the convolution of R-Ri series with the rectangular window used in their method has the effect on the power spectrum of the multiplication by a low-pass filter. They described the shape of the filter as:

$$W(f) = \left[\frac{\sin\left(2\pi f / f_s\right)}{2\pi f / f_s} \right]^2 \qquad (7)$$

where f_s is the sampling frequency of the heart rate signal and $2/f_s$ is the width of the rectangular window in the time domain.

The filter exerts a negligible power beyond the Nyquist rate (i.e., $f_s/2$) and the effect could be compensated in the band $0 < f < f_s/2$ multiplying the power spectrum by $1/W(f)$. This correction applied to the spectrum significantly amplifies the aliased power in the band $f_s/4 < f < f_s/2$, so that the accurate spectral estimate is limited for $0 < f < f_s/4$. While using 4 Hz as the sampling frequency, it enables the computation of reliable PSD estimates between DC and 1 Hz.

Laguna et al. (1998) have also described a low-pass filtering effect introduced by resampling, and compared the use of a FFT method with the Lomb-Scargle method (Scargle 1982; Lomb 1976) that may evaluate the PSD of non-uniformly sampled in time data sequences, avoiding a low-pass filtering effect. A low-pass effect on the power spectrum of resampling consists of an attenuation of high frequencies. The use of a cubic spline interpolation approaches the results of Lomb's estimates and suggests that the resampling should be done at least to double the mean sampling frequency, to avoid aliasing, which for heart rate signals has been fixed at about 2 Hz.

In the present study we applied a resampling frequency of 4 Hz, and the method used for the interpolation was the cubic spline. Thus, we reasonably expected that a frequency range of 0.4–0.9 Hz, being inside the frequency range of $0 < f < f_s/4$, could be reliably explored considering methodological issues. A comparative analysis of the spectral indices, calculated with the sampling interpolation frequencies of 2 or 4 Hz, showed a low-pass filtering effect, consisting of a relative energy reduction in the higher frequency bands of the spectra (LF, HF, and VHF). There also was an increase in VLF energy content when using a lower resampling frequency of 2 Hz, but the effect was only detected by the novel Spa-EMD method and not by classical Spa-Welch method. This effect was similar for both healthy and patient groups. These results point out the inconvenience of using low values of resampling frequency, which could be seen when using a novel spectral approach of Spa-EMD, developed for use even in nonlinear and non-stationary time-series analysis (Huang et al. 1998). A classical spectral Spa-Welch method failed to detect a low-pass filtering effect of the resampling process in the present study.

Advantages and limitations of the Lomb-Scargle algorithm in spectral analysis of HRV have been discussed in recent papers of Estévez-Báez et al. (2016a, b). The most important limitation of this approach could be related to the Nyquist rate (fc). Suppose we have the HR of 120 beats/min. This frequency expressed in Hz would be 2 Hz and the corresponding sampling period would be 0.5 s. The fc is calculated as [1/(2 * sampling period)], so that the fc would be [1/ (2 * 0.5)] = 1 Hz. But if the HR were 70 beats/min, applying the same calculations, the fc would be of approximately 0.59 Hz. In other words, as we were interested in a frequency range of 0.4–0.9 Hz, the Lomb-Scargle algorithm would not be a good choice.

4.4　Origin of Very High Frequency (VHF) Components of Heart Rate Variability (HRV)

Although only a few investigations have reported the existence of VHF oscillations (> 0.4 and <1 Hz) in HRV, most hypothesize about the origin of these components. The presence of peaks of very low amplitude, but clearly in the VHF range in some patients after orthotopic cardiac transplantation during exercise is suggestive of intrinsic activity of the heart muscle *via* atrial stretch (Bernardi et al. 1989, 1990). Studies carried out in cardiac transplant patients have shown the presence of VHF components (Toledo et al. 2003), which is considered as being related to the absence of a significant feedback of a parasympathetic influence on the heart (Spallone et al. 2011). The presence of VHF activity is thought to help identify the lack of a parasympathetic heart control, for instance, in cases of brain death, neurodegenerative diseases, or diabetes. Such studies include a spectral assessment of the simultaneously recorded beat-to-beat blood pressure that also displays the VHF components, which are not visible in the respiratory signal.

Studies in patients with ischemic heart disease, performing exercise testing, have ascribed the origin of VHF components to the sympathetic tone, since these components are best detected during cardiovascular stress (Bailon et al. 2003; Mateo et al. 2001). Other authors have used the VHF, emphasizing that its origin is unclear, but it shows a clear potential to differentiate the dynamics to different autonomic maneuvers in healthy subjects (Chang et al. 2014a, b) and between healthy subjects and patients with cardiac disorders, including children (Özyιlmaz et al. 2015; Xia 2009).

No investigation has ever been carried out design specifically to elucidate the origin of this frequency range in HRV, which remains hitherto conjectural. The present study did not address the origin of VHF either. The present comparative analysis of VHF indices suggests that the activity in this frequency range may be related to the intensity of a reduction in the cardiac parasympathetic influence as a consequence of autonomic neural fiber damage, which is present in CAN, with a possible involvement of the intrinsic autonomic nervous system of the heart. To the best of our knowledge, the present report is the first which explores the VHF components in patients diagnosed with cardiovascular autonomic neuropathy, a severe type of autonomic neuropathy underlain by different etiologies.

5 Conclusions

We submit that VHF activity is not an artefactual phenomenon and that it can be identified in control healthy humans. The observed VHF changes are attributable to severe vagal denervation produced by cardiac autonomic neuropathy in SCA2 patients. However, we cannot discard neural autonomic denervation of sympathetic nerve fibers, quantitatively exposed in the present study. The present study demonstrates that the examination of the VHF band of HRV seems a promising tool in the care of patients with reduced central autonomic control of the cardiovascular system, in particular of the heart. It would be interesting to explore the VHF indices in types of CAN and in comatose patients with different degrees of impaired consciousness, and in patients in brain death, the conditions in which the parasympathetic vagal innervation is functionally or pathologically damaged, which is conducive to the presence of VHF activity.

Acknowledgments The authors' thanks go to the Laboratory of Signals and Nonlinear Dynamics of the University of Entre Ríos in the Republic of Argentina, especially to Marcelo A. Colominas and Profs. Gastón Schlotthauer and María E. Torres for their splendid work in the development and modifications of the CEEMDAN methods, with the corresponding implementations in Matlab, downloaded from https: www.bioingenieria.edu.ar/grupos/ldnlys for use in the present study.

Competing Interests The authors declare that they have no competing interests in relation to this article.

References

Alonso A, Huang X, Mosley TH, Heiss Gand Chen H (2015) Heart rate variability and the risk of Parkinson's disease: the Atherosclerosis Risk in Communities (ARIC) study. Ann Neurol 77:877–883

Baevskii RM, Zamotaev IP, Nidekker IG (1971) Mathematical analysis of sinus automatism for prognosis of rhythm disorders. Kardiologiia 11:65–68

Bailon R, Mateo J, Olmos S, Serrano P, Garcia J, del Rio A, Ferreira IJ, Laguna P (2003) Coronary artery disease diagnosis based on exercise electrocardiogram indexes from repolarisation, depolarisation and heart rate variability. Med Biol Eng Comput 41:561–571

Balcioglu AS, Muderrisoglu H (2015) Diabetes and cardiac autonomic neuropathy: clinical manifestations, cardiovascular consequences, diagnosis and treatment. World J Diabetes 6:80–91

Berger RD, Akselrod S, Gordon D, Cohen RJ (1986) An efficient algorithm for spectral analysis of heart rate variability. IEEE Trans Biomed Eng 33:900–904

Bernardi L, Keller F, Sanders M, Reddy PS, Griffith B, Meno F, Pinsky MR (1989) Respiratory sinus arrhythmia in the denervated human heart. J Appl Physiol 67:1447–1455

Bernardi L, Salvucci F, Suardi R, Solda PL, Calciati A, Perlini S, Falcone C, Ricciardi L (1990) Evidence for an intrinsic mechanism regulating heart rate variability in the transplanted and the intact heart during submaximal dynamic exercise? Cardiovasc Res 24:969–981

Boulton AJM, Vinik AI, Arrezzo JC, Bril V, Feldman EL, Freeman R, Malik R, Maser R, Sosenko JM, Ziegler D (2005) Diabetic neuropathies. A statement by the American Diabetes Association. Diabetes Care 28:956–962

Chang CC, Hsiao TC, Hsu HY (2014a) Frequency range extension of spectral analysis of pulse rate variability based on Hilbert-Huang transform. Med Biol Eng Comput 52:343–351

Chang CC, Hsu HY, Hsiao TC (2014b) The interpretation of very high frequency band of instantaneous pulse rate variability during paced respiration. Biomed Eng Online 13:46

Chyun DA, Wackers FJ, Inzucchi SE, Jose P, Weiss C, Davey JA, Heller GV, Iskandrian AE, Young LH, DIAD Investigators (2015) Autonomic dysfunction independently predicts poor cardiovascular outcomes in asymptomatic individuals with type 2 diabetes in the DIAD study. SAGE Open Med 3:2050312114568476

Colominas MA, Schlotthauer G, Torres ME (2014) Improved complete ensemble EMD: a suitable tool for biomedical signal processing. Biomed Signal Process Control 14:19–29

Cygankiewicz I, Zareba W (2013) Heart rate variability. Handb Clin Neurol 117:379–393

DeBoer RW, Karemaker JM, Strackee J (1984) Comparing spectra of a series of point events particularly for heart rate variability data. IEEE Trans Biomed Eng 31:384–387

Estévez-Báez M, Machado C, Leisman G, Brown-Martínez M, Jas-García JD, Montes-Brown J, Machado-García A, Carricarte-Naranjo C (2016a) A procedure to correct the effect of heart rate on heart rate variability indices: description and assessment. Int J Disabil Hum Dev. https://doi.org/10.1515/ijdhd-2015-0014

Estévez-Báez M, Machado C, Leisman G, Estévez-Hernández T, Arias-Morales A, Machado A, Montes-Brown J (2016b) Spectral analysis of heart rate variability. Int J Disabil Hum Dev. https://doi.org/10.1515/ijdhd-2014-0025

Ewing DJ, Martyn CN, Young RJ, Clarke BF (1985) The value of cardiovascular autonomic function tests: 10 years experience in diabetes. Diabetes Care 8:491–498

Huang NE, Shen Z, Long SR, Wu MC, Shih HH, Zheng Q, Yen N, Tung CC, Liu HH (1998) The empirical mode decomposition and the Hilbert spectrum for nonlinear and non-stationary time series analysis. Proc R Soc Lond 454:903–995

José AD, Collison D (1970) The normal range and determinants of the intrinsic heart rate in man. Cardiovasc Res 4:160–167

Kuusela T (2013) Methodological aspects of heart rate variability analysis. In: Kamath MV (ed) Heart Rate Variability (HRV) signal analysis: clinical applications. CRC Press, Boca Raton

Lacigova S, Brozova J, Cechurova D, Tomesova J, Krcma M, Rusavy Z (2016) The influence of cardiovascular autonomic neuropathy on mortality in type 1 diabetic patients; 10-year follow-up. Biomed Pap Med Fac Univ Palacky Olomouc Czech Repub 160:111–117

Laguna P, Moody GB, Mark RG (1998) Power spectral density of unevenly sampled data by least-square analysis: performance and application to heart rate signals. IEEE Trans Biomed Eng 45:698–715

Lomb NR (1976) Least-squares frequency analysis of unequally spaced data. Astrophys Space Sci 39:447–462

MacFarlane PW, Van Oosterom A, Pahlm O, Kligfield P, Janse M, Camm J (2010) Comprehensive electrocardiology. Springer, New York

Machado C, Estévez M, Pérez-Nellar J, Gutiérrez J, Rodríguez R, Carballo M, Chinchilla M, Machado A, Portela L, García-Roca MC, Beltrán C (2011) Autonomic, EEG, and behavioral arousal signs in a PVS case after zolpidem intake. Can J Neurol Sci 38:341–344

Machado C, Estévez M, Rodriguez R, Perez-Nellar J, Chinchilla M, DeFina P, Leisman G, Carrick FR,

Melillo R, Schiavi A, Gutierrez J, Carballo M, Machado A, Olivares A, Perez-Cruz N (2014) Zolpidem arousing effect in persistent vegetative state patients: autonomic, EEG and behavioral assessment. Curr Pharmaceut Des 20:4185–4202

Machado-Ferrer Y, Estévez M, Machado C, Hernandez-Cruz A, Carrick FR, Leisman G, Melillo R, Defina P, Chinchilla M, Machado Y (2013) Heart rate variability for assessing comatose patients with different Glasgow Coma Scale scores. Clin Neurophysiol 124:589–597

Marple LS (1999) Computing the discrete-time 'analytic' signal via FFT. IEEE Trans Signal Process 47:2600–2603

Mateo J, Serrano P, Bailón R, Olmos S, García J, del Río A, Ferreira I, Laguna P (2001) ECG-based clinical indexes during exercise test including repolarization, depolarization and HRV. Comput Cardiol 28:309–312

Middleton PM, Chan GSH, Marr S, Celler BG, Lovell NH (2011) Identification of high-risk acute coronary syndromes by spectral analysis of ear photoplethysmographic waveform variability. Physiol Meas 32:1181–1192

Montes-Brown J, Sanchez-Cruz G, Garcia AM, Estévez-Báez M, Velazquez-Perez L (2010) Heart rate variability in type 2 spinocerebellar ataxia. Acta Neurol Scand 122:329–335

Montes-Brown J, Machado A, Estévez M, Carricarte C, Velazquez-Perez L (2012) Autonomic dysfunction in presymptomatic spinocerebellar ataxia type-2. Acta Neurol Scand 125:24–29

Mukaino A, Nakane S, Higuchi O, Nakamura H, Miyagi T, Shiroma K, Tokashiki T, Fuseya Y, Ochi K, Umeda M, Nakazato T, Akioka S, Maruoka H, Hayashi M, Igarashi SI, Yokoi K, Maeda Y, Sakai W, Matsuo H, Kawakami A (2016) Insights from the ganglionic acetylcholine receptor autoantibodies in patients with Sjogren's syndrome. Mod Rheumatol 26:708–715

Nidekker IG (1981) Method of spectral analysis for long-term recordings of physiological curves. Kosm Biol Aviakosm Med 15:78–82

Özyılmaz I, Ergül Y, Tola HT, Saygı M, Öztürk E, Tanidir İC, Tosun Ö, Özyılmaz S, Gül M, Güzeltaş A, Ödemiş E (2015) Heart rate variability improvement in children using transcatheter atrial septal defect closure. Anatol J Cardiol 16:290–295

Parin VV, Baevskii RM, Gazenko OG (1965a) Achievements and problems of modern space cardiology. Kardiologiia 22:3–11

Parin VV, Baevskii RM, Gazenko OG (1965b) Heart and circulation under space conditions. Cor Vasa 32:165–184

Pinhas I, Toledo E, Aravot D, Akselrod S (2004) Bicoherence analysis of new cardiovascular spectral components observed in heart-transplant patients: statistical approach for bicoherence thresholding. IEEE Trans Biomed Eng 51:1774–1783

Piskorski J, Guzik P, Krauze T, Zurek S (2010) Cardiopulmonary resonance at 0.1 Hz demonstrated by averaged Lomb-Scargle periodogram. Centr Eur JPhys 8:386–392

Pradhan C, Yashavantha BS, Pal PK, Sathyaprabha TN (2007) Spinocerebellar ataxias type 1, 2 and 3: a study of heart rate variability. Acta Neurol Scand 117:337–342

Rasheedy D, Taha HM (2016) Cardiac autonomic neuropathy: The hidden cardiovascular comorbidity in elderly patients with chronic obstructive pulmonary disease attending primary care settings. Geriatr Gerontol Int 16:329–335

Sassi R, Cerutti S, Lombardi F, Malik M, Huikuri MH, Peng CK, Schmidt G, Yamamoto Y (2015) Advances in heart rate variability signal analysis: joint position statement by the e-Cardiology ESC Working Group and the European Heart Rhythm Association co-endorsed by the Asia Pacific Heart Rhythm Society. Europace 17:1341–1353

Scargle JD (1982) Studies in astronomical time series analysis. II. Statistical aspects of spectral analysis of unevenly spaced data. Astrophys J 263:835–853

Spallone V, Ziegler D, Roy F, Bernardi L, Frontoni S, Pop-Busui R, Stevens M, Kempler P, Hilsted J, Tesfaye S, Low P, Valensi P (2011) Cardiovascular autonomic neuropathy in diabetes: clinical impact, assessment, diagnosis, and management. Diabet Metab Res Rev 27:639–653

Strobel JS, Epstein AE, Bourge RC, Kirklin JK, Kay GN (1999) Nonpharmacologic validation of the intrinsic heart rate in cardiac transplant recipients. J Interv Card Electrophysiol 3:15–18

Task Force of ESC and NASPE (1996) Heart rate variability. Standards of measurement, physiological interpretation, and clinical use. Task Force of the European Society of Cardiology and the North American Society of Pacing and Electrophysiology. Eur Heart J 17:354–381

Toledo E, Pinhas I, Aravot D, Akselrod S (2003) Very high frequency oscillations in the heart rate and blood pressure of heart transplant patients. Med Biol Eng Comput 41:432–438

Torres ME, Colominas MA, Schlotthauer G, Flandrin P (2011) A complete ensemble empirical mode decomposition with adaptive noise. In: 36th international conference on Acoustics, Speech and Signal Processing ICASSP, Prague, Czech Republic

Vinik AI, Erbas T (2013) Diabetic autonomic neuropathy. Handb Clin Neurol 117:279–294

Vinik AI, Ziegler D (2007) Diabetic cardiovascular autonomic neuropathy. Circulation 115:387–397

Vinik AI, Erbas T, Pfeifer M, Feldman E, Stevens M, Russell J (2004) Diabetic autonomic neuropathy. In: Inzucchi SE (ed) The diabetes mellitus manual: a primary care companion to Ellenberg and Rifkin's 6th edn. McGraw Hill, New York, p 351

Vinik AI, Erbas T, Casellini CM (2013) Diabetic cardiac autonomic neuropathy, inflammation and cardiovascular disease. J Diabet Invest 4:4–18

Wheelock KM, Jaiswal M, Martin CL, Fufaa GD, Weil EJ, Lemley KV, Yee B, Feldman E, Brosius FC 3rd, Knowler WC, Nelson RG, Pop-Busui R (2016) Cardiovascular autonomic neuropathy associates with nephropathy lesions in American Indians with type 2 diabetes. J Diabet Complications 30:873–879

Widyanti A, de Waard D, Johnson A, Mulder B (2013) National culture moderates the influence of mental effort on subjective and cardiovascular measures. Ergonomics 56:182–194

Xia L (2009) A very high frequency index of heart rate variability for evaluation of left ventricular systolic function and prognosis in chronic heart failure patients using five-minute electrocardiogram. J Geriatr Cardiol 6:213–217

Xu W, Zhu Y, Yang X, Deng H, Yan J, Lin S, Yang H, Chen H, Weng J (2016) Glycemic variability is an important risk factor for cardiovascular autonomic neuropathy in newly diagnosed type 2 diabetic patients. Int J Cardiol 215:263–268

Zhang Y, Chan GS, Tracy MB, Lee QY, Hinder M, Savkin AV, Lovell NH (2011) Spectral analysis of systemic and cerebral cardiovascular variabilities in preterm infants: relationship with clinical risk index for babies (CRIB). Physiol Meas 32:1913–1928

Zhemaitite D (1967) Methods of mathematical analysis of the heart rhythm. Nauka, Moskow

Zoppini G, Cacciatori V, Raimondo D, Gemma M, Trombetta M, Dauriz M, Brangani C, Pichiri I, Negri C, Stoico V, Bergamini C, Targher G, Santi L, Thomaseth K, Bellavere F, Bonadonna RC, Bonora E (2015) Prevalence of cardiovascular autonomic neuropathy in a cohort of patients with newly diagnosed type 2 diabetes: The Verona newly diagnosed type 2 diabetes study (VNDS). Diabetes Care 38:1487–1493

Advs Exp. Medicine, Biology - Neuroscience and Respiration (2018) 39: 71–84
https://doi.org/10.1007/5584_2018_151
© Springer International Publishing AG 2018
Published online: 22 March 2018

Improvement in Hand Trajectory of Reaching Movements by Error-Augmentation

Sharon Israely, Gerry Leisman, and Eli Carmeli

Abstract

The purpose of this study was to investigate whether adaptive responses to error-augmentation force fields, would decrease the trajectory errors in hand-reaching movements in multiple directions in healthy individuals. The study was conducted, as a randomized controlled trial, in 41 healthy subjects. The study group trained on a 3D robotic system, applying error-augmenting forces on the hand during the execution of tasks. The control group carried out the same protocol in null-field conditions. A mixed-model ANOVA was implemented to investigate the interaction between groups and time, and changes in outcome measures within groups. The findings were that there was a significant interaction effect for group × time in terms of the magnitude of movement errors across game-sets. The trajectory error of the study group significantly decreased from 0.035 ± 0.013 m at baseline to 0.029 ± 0.011 m at a follow-up, which amounted to a 14.8% improvement. The degree of movement errors were not significantly changed within a game-set. We conclude that practicing hand-reaching movement in multiple random directions, using the error-augmentation technique, decreases the deviation of the hand trajectory from a straight line. However, this type of training prevents the generalizability of adaptation between consecutive reaching movements. Further studies should investigate the feasibility of this training method for rehabilitation of post-stroke individuals.

Keywords

Adaptation · Brain model · Error-augmentation · Force-field · Hand reaching · Hand trajectory

1 Introduction

A fundamental concept in motor learning is the model-based paradigm, which suggests that motor learning is driven by previous movement experiences. According to this paradigm, unexpected external perturbation forces acting on the arm, will deflect the arm's trajectory from the preplanned path, causing reaching errors. Repetitive execution of the same reaching movement, under the same systematic forces, results in a gradual

S. Israely (✉) and E. Carmeli
Department of Physical Therapy, University of Haifa, Haifa, Israel
e-mail: sharonis@mh.org.il

G. Leisman
Department of Physical Therapy, University of Haifa, Haifa, Israel

The National Institute for Brain and Rehabilitation Sciences-Israel, Nazareth, Israel

decrease in errors, until the same motor performance as before is achieved (Huang et al. 2011).

The process of reducing the movement errors by adaptation to external forces, likely occurs due to recurrently updating a preplanned motor program, i.e., internal model (Izawa et al. 2012). The internal model predicts the consequences of the next motor command according to the dynamics of the environment. Therefore, in each movement the afferent sensory messages calibrate their internal model according to the previous movement.

Numerous studies have investigated the adaptation capacities in hand reaching tasks, but also in balancing and gait both in healthy subjects and in individuals with neurological disorders. In healthy subjects, for example, it has been reported that error augmentation force fields increase the accuracy of hand trajectory during curve-tracing tasks compared to error minimization force field (Williams et al. 2016). Augmenting movement errors during practice have also been applied to improve symmetry of gait post-stroke (Bishop et al. 2017; Lewek et al. 2017), spinal cord injury (Yen et al. 2014) and in standing in healthy (O'Brien et al. 2017), but the results are inconsistent, compared to the application of minimizing movement errors.

Other studies have confirmed the expansion of the model-based paradigm, to a state-space model, which suggests that error-based learning mechanisms have the capacity to be generalized to movements in other directions and different trajectories (Huang et al. 2011; Donchin et al. 2003). In this context, if an internal model represents mapping the arm's position and velocity into force, then any change in the position or the velocity will be reflected by changes in the resultant force. In other words, according to this model, changing the direction and range of motion should not significantly impact the rate of adaptation, as observed in previous studies (Shadmehr and Moussavi 2000). However, some of the studies that have investigated the generalizability of the adaptation process, dealt with reaching movements only in the horizontal plain, with arm support in which movement lengths were 10 cm (Conditt et al. 1997; Donchin et al. 2003) or 20 cm (Sainburg et al. 1999).

From the perspective of neurorehabilitation, the augmented force fields are designated to use the adaptation properties of the central nervous system (CNS) to induce normal movement patterns after a stroke by error-augmentation (EA) training (Emken and Reinkensmeyer 2005; Patton et al. 2006a). Rehabilitation training applied by EA force fields is based on the notion that movement errors drive learning (Patton et al. 2006b), increase the signal to noise ratio, and make movement error more perceptible, thus enhancing movement correction. However, a recent systematic review of studies that implement EA training on the upper extremity of stroke patients has provided only partial support for the effectiveness of this training concept over null-field robotic training (Israely and Carmeli 2016; Huang and Patton 2013; Rozario et al. 2009).

Summarizing all of the above references, it seems that there is scarce evidence regarding the generalizability properties of adaptation in diversity of movement lengths and directions and in the three dimensional space. Since hand reaching movements are usually executed in 3D space without arm support and involve full ranges of movements of the shoulder and elbow, given the generalizability properties of adaptation, it is crucial to test the adaptation paradigm under these conditions. This is especially relevant when considering the use of adaptation paradigm for rehabilitation purposes after CNS lesions in order to enhance normal movement patterns. Therefore, based on the current knowledge in motor learning, it is reasonable to assume that training under augmented environment would enhance adaptation and would reduce trajectory errors.

In this study, we investigated the adaptive trajectory responses to EA force fields training in hand reaching movements for multiple directions in healthy individuals. The EA algorithm was tailored specifically to the length of the arm, and the power of the arm, and the participant's current level of performance. The forces were computed in real-time, based on the current distance of the hand from the straight trajectory line and adjusted to the phase of the movement while making the reaching task. We hypothesized that EA training would result in decreased trajectory errors compared to standard robotic training.

2 Methods

2.1 Participants

Forty-one participants (15 males and 26 females), were recruited and were randomly assigned to one of the two training groups. Each participant provided informed consent in accordance with the University of Haifa Institutional Review Board requirements. Participants ranged in age from 20 to 50 (mean age 25 ± 10SD). Participants' characteristics are summarized in Table 1.

2.2 Apparatus

We used the 3D DeXtreme robotic system (BioXtreme Ltd.; Rechovol, Israel) (Fig. 1). The base of the apparatus is 870 mm (height)× 600 mm × 600 mm. The robotic arm is attached to an axis inside the base of apparatus. It has two segments: one that is attached to the base, which is 500 mm long, and the second segment attached to the first one at one side and to a gimbal handle at the end of it, which is 400 mm long. The weight of the robotic arm is 45 kg. The DeXtreme is capable of generating transient peaks of 100 N. However, for safety reasons the device was programed to allow no more than 40 N. The maximal inertia (P) at the end of the robotic arm is defined by $p\left(\frac{kg \times m}{sec}\right) = 5 \times V$, and therefore resulted in $2.5\left(\frac{kg \times m}{sec}\right)$ at maximal hand velocity of 0.5 m/sec. The robotic motors were programed to eliminate the weight of the robotic arm either in the null-field environment or with error-augmenting forces. Three sensors tracked the hand position every 20 msec. The device was equipped with a 25 inch. LCD screen, on which the motor task was displayed.

In the horizontal plane the workspace was a trapezoid shape aligned according to Fig. 2. The vertical axis of the workspace allowed a range of motion of about 360 mm. Accordingly, the range of reaching movements during the training sessions were between 130 mm and 360 mm.

Hand reaching movements were carried out in a three dimensional workspace as illustrated in Fig. 3. The robotic system raised the difficulty of the hand reaching task according to previous successful attempts by changing the length of the trajectory path and increasing the reaching angles to the sides of the hand-reaching space. Although increasing the difficulty of reaching movements may increase the variability between participants, it is crucial for intensive workout necessary for motor learning post CNS lesion (Krakauer et al. 2012). We assumed that including twenty participants in each group, with each individual performing movements in approximately 100 random directions, would obviate the effect of movement direction as a confounding variable. The extent of perturbation forces was constant independent of the difficulty level. Figure 3 illustrates an example of 16 trajectory vectors, expressing the order of 16 straight trajectory lines of a single game set. Each four consecutive movement arrows in Fig. 3 were similarly colored to exemplify the randomization of movement direction and length. A successful movement was considered to be completed within 2 s.

The starting point for each reaching movement for all participants was standardized by calibrating the participants to the device workspace. Nevertheless, Fig. 3 illustrates some variations in the starting points, probably due to

Table 1 Baseline characteristics of groups

	Study ($n = 20$)	Control ($n = 21$)	p-value
Gender			0.393
Male	6	9	
Female	14	12	
Age (years)	31.8 ± 8.3	33.2 ± 6.8	0.573
Dominant hand			
Right	20	17	0.040
Left	0	4	
Baseline trajectory error (m)	0.034 ± 0.012	0.029 ± 0.011	0.177

Data are means ±SD

Fig. 1 Illustration of the DeXtreme robotic device

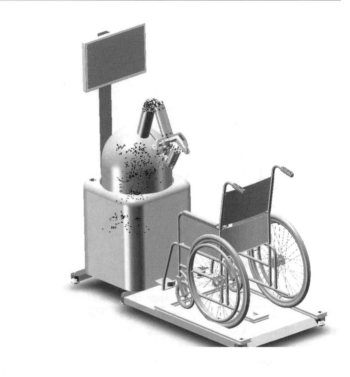

Fig. 2 2D upward view of the experimental horizontal workspace of the DeXtreme Robotic interface

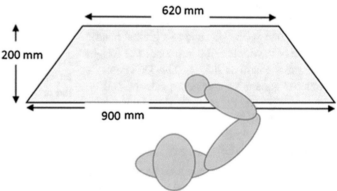

technical inaccuracies. The starting point for each participant involved slightly different joint positions. Generally speaking, the upper extremity at the starting position was approximately 20° shoulder flexion, with comfortable shoulder internal rotation and forearm supination. The system's control mechanism is illustrated in the flow diagram of Fig. 4, summarizing the integration between human and the robotic device.

2.3 Force Field Algorithm

Error-augmenting forces F_{Total} were applied by the robotic apparatus in a direction of forces distant from the straight trajectory line, and perpendicular to the straight trajectory line. It was calculated by integrating two functions: 1) F_{cal} calculates the forces that should be applied on the hand, as a function of the distance of the hand from the straight trajectory line and 2) F_{ROM} calculates the forces that should be applied on the hand that depend on the path that the hand had reached relative to the complete trajectory of the same movement. The F_{ROM} was applied to compensate for increased shoulder torques and changing dynamics as the hand moved away from the body, which previously was observed as a 'hooking' phenomena in a 'curl force-field' (Shadmehr and Brashers-Kurg 1997; Shadmehr and Mussa-Ivaldi 1994). In other

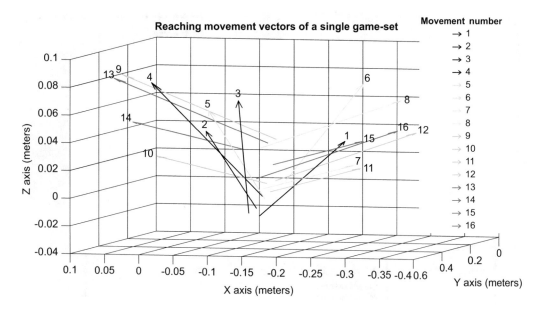

Fig. 3 Directional vectors of the reaching movements of a single game-set. The order of reaching movements were determined according to the motor performance in previous movements. The difficulty level reaching movement was defined by the length and the direction of the movement. In each practice level the order of the movements were randomly selected by the DeXtreme robotic device. Technical issues were imposed a different the X and Z starting position of a hand reaching task, but not the Y axis were different within a game set

words, the same magnitude of force might be associated with increased burden on the shoulder musculature when the extremity is fully extended at the end stage of reaching, than when the hand is close to the body at the beginning of the reaching movement. The F_{ROM} was applied to decrease this phenomenon, by decreasing the forces as the arm extended away from the body. Schematic illustration of F_{cal} and F_{ROM} is demonstrated in Fig. 5. The EA force functions were calculated according to Eq. 1, where, F_{max} is a constant that was determined before the beginning of a training session. In order to normalize the maximal forces applied on the hand across participants (F_{max}), a maximal isometric contraction test of shoulder horizontal adduction (F_{Test}) was carried out three times, and averaged. This was followed by further dividing the F_{Test} value by three, based on our previous experience to allow significant adaptation while still providing safe and comfortable interaction (Givon-Mayo et al. 2014). Accordingly the F_{max} was calculated according to Eq. 2. F_{ROM} was calculated according to Eq. 3 and F_{Total} was calculated according to Eq. 4.

$$F_{cal} = \begin{cases} 0 & if & E_{cal} \leq E_{Fmin} \\ \dfrac{E_{cal} - E_{Fmin}}{E_{Fmax} - E_{Fmin}} \times F_{max} & if & E_{Fmin} < E_{cal} \leq E_{Fmax} \\ F_{max} & if & E_{FM} < E_{cal} \leq E_D \end{cases} \tag{1}$$

$$F_{max} = F_{Test} \times 1/3 \tag{2}$$

$$F_{ROM} = F_{max} - \left[\frac{P_{ROM}}{100\%} \times (F_{max} - F_{Const}) \right] \tag{3}$$

$$F_{Total} = F_{cal} \times F_{ROM}/F_{max} \tag{4}$$

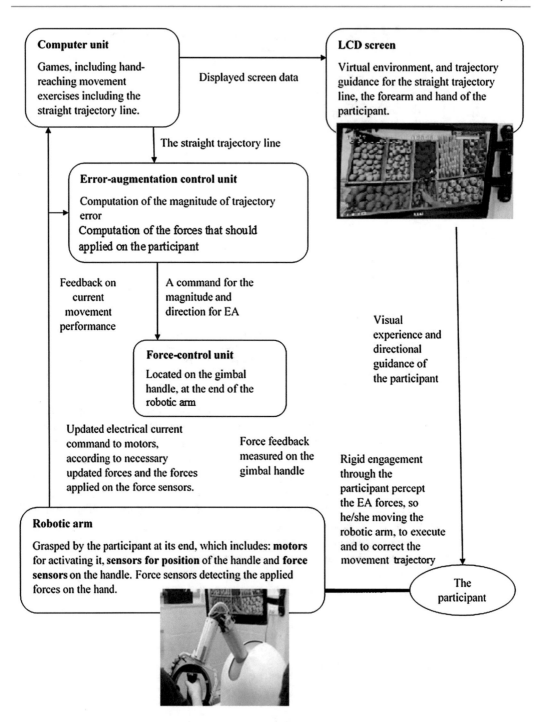

Fig. 4 Flow chart of the control mechanism. The game application consisted of a virtual colorful vegetable market stand and an avatar hand representing the hand of the participant, which are displayed in the upper right-hand corner of the chart. A virtual bee moved to a starting point at the bottom of the screen. The participant was required to bring the avatar hand to the starting point. When the avatar hand reached the starting point, the bee moved to a random location on the screen. The participant was required to use a hand to catch the bee. Successful attempts were displayed as a flare, while non-successful attempt caused the bee to move from its location. The DeXtreme devise used for the game-set is displayed in the bottom of the chart

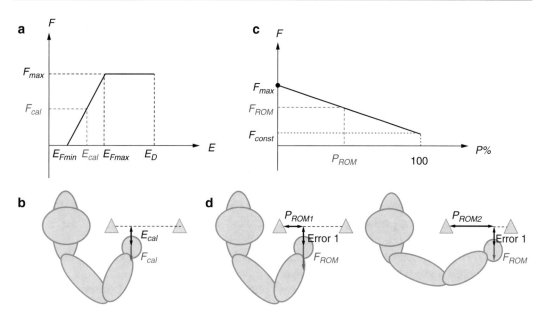

Fig. 5 Force-field algorithm. (**a**) The error augmenting forces were calculated as a function of the distance of the hand from the straight trajectory line. The Y axis refers to the forces that were applied on the hand, and the X axis refers to the error from the straight trajectory line. F_{max} is the maximal force that could be applied on the hand; F_{cal} is the calculated force given the trajectory error E_{cal}. E_{Fmin} is the maximal trajectory error without application of forces; E_{Fmax} is the degree of trajectory error in which maximal forces applied; E_D is the maximal trajectory error in which the forces applied. Red dashed line illustrates an example of current position of a hand relative to the straight trajectory line (E_{cal}), and its respective calculated forces which will be applied on the hand (F_{cal}). (**b**) During hand-reaching movement, the trajectory error (E_{cal}) was measured as the shortest distance of the hand from the straight trajectory line. The robotic device calculated the perturbation force (F_{cal}) that was directed further away from the straight trajectory line, and perpendicular to it. (**c**) The F_{ROM} function calculates the forces as a function of the the current position of the hand in percentage from the total range of motion of the specific hand reaching movement. Red dashed lines illustrates the current position of the hand (P_{ROM}) and the corresponding calculated force F_{ROM}. (**d**) Illustration of the F_{ROM} function. The F_{ROM} decreases as the arm extended away from the body, independent of a constant error. The algorithm further integrated the F_{cal} with the F_{ROM} for calculating the resultant force applied on the hand F_{Total}

2.4 Study Procedure

Participants were randomly allocated to one of the two groups by using sealed opaque envelopes (Doig and Simpson 2005). Each participant received either a control treatment of repetitive practice with no EA, called the control treatment, or a treatment with the same amount of practice plus haptic EA, called the error augmentation treatment (EA). For both groups, the treatment protocol included practicing of arm reaching movements for multiple directions in three dimensional space.

Before beginning the protocol, the operator entered the personal data for the computation of forces applied during the practice trials. The training workspace was calibrated according to the length of each participant's arm. For both groups the session began with a 2-min game, identical to the practicing games, which is described in detail below. This session was without perturbation forces, so that the participant could become accustomed to the device. This was followed by another game of the same length for the baseline assessment. This, in turn, was followed by five games of 2 min each according to the group allocation. Participants were allowed to rest for 1–2 min between the games. Participants then carried out another follow-up game without forces.

2.5   Treatment Session Protocol

During the practice sessions, participants were comfortably seated in a chair and held the robotic handle while viewing a visual screen. The game application consisted of a virtual vegetable market stand, an avatar hand representing the hand of the participant and a bee that was used as a reaching target (Fig. 4). By moving the robotic arm, the participants were able to see the moving avatar hand on the screen. A flying bee moved to a starting point at the bottom of the screen and stopped at this point. The participant brought his or her arm to the starting point where the bee was located. When the hand of the participant reached the starting point, the bee moved fast on a straight line, which was indicated on-screen, to a random point on the screen and stopped. Therefore, the participant could see the whole path of the bee as it moves on the straight line toward the final location, as well as the actual straight line of its path. The participant could also see the straight trajectory during the execution of the reaching movement.

The participant executed a hand-reaching movement within 2 s in order to place the avatar hand on the bee. This was then considered to be a successful attempt. Upon striking the bee by the avatar hand, a flare was displayed and the bee and the straight line disappeared. Then, another bee moved to the starting point for the next attempt. In a successful attempt, the time from the end of a movement to the beginning of next one was 4 s. Therefore each 2-min game consisted of about 20 reaching movements according to the participant's performance. On-screen visual feedback was also displayed, indicating the running time of the game, the number of successful trials and weighted scores of the movement-error and successful attempts.

2.6   Outcome Measures

Two measures were used to analyze the data, either was calculated for the complete hand reaching movements or just the first 300 msec of

movement. Dividing the task into two separated stages was designated to distinguish between to control mechanisms: update of internal model and feedback mechanism. Previous studies reported that the first 300 msec of a movement are not affected by feedback mechanisms that enables to correct the hand trajectory during the task execution (Patton et al. 2006a).

For each reaching movement at each time point the shortest distance between the location of the hand and the closest point on the straight hand trajectory line was calculated. This was applied by iteratively calculating the distance between a point (hand location) and a line (provided by the robotic device) in 3 dimensions.

2.6.1   Mean Movement Error

The trajectory errors of all movements within a game-set were averaged, resulting in a single scalar value representing the average trajectory error of a game-set. Therefore, each participant received seven values for the whole session. The mean movement error was calculated separately for the complete movements, and the first 300 msec of the reaching movements.

2.6.2   Movement Error Within a Game-Set

The trajectory error was computed for each reaching movement during a game-set and normalized to the length of the movement to eliminate the influence of movement length on the magnitude of the trajectory error. In contrast, this measure disregarded the inherent differences that might be concealed between different movement directions. Movement errors were calculated for each movement in each of the seven game-sets and averaged for the whole session of a particular participant.

2.7   Statistical Analysis

The hand-trajectory row data were measured and recorded by the robotic system and were later processed in MATLAB R2016B (MathWorks; Natick, MA). A mixed-design with repeated measures ANOVA was carried out to test the interactions and the main effects for time and groups throughout the study, both for complete reaching movement and the initial phase of 300 msec. We initially carried out the 2 × 2 mixed model ANOVA using the baseline game-set and follow-up game-set measurements as within-group factors. The group allocation was used as a between-subject factor. This was followed by a pairwise comparisons using the Bonferroni method to further study the effect of intervention for each group separately between time points, and the differences between groups at each time point. Based on the initial results we add the baseline scores as a covariant factor. The same statistical methods were also applied to analyze the magnitude of errors within a game-set. The α-level of 0.05 was considered to be statistically significant. Statistical evaluation was performed in SPSS 21 (IBM; Armonk, NY).

3 Results

No significant differences were revealed at baseline between the two groups (Table 1). The mean calculated F_{max} in the study group were 30.59 ± 7.09 N and 16.65 ± 4.17 N for males and females, respectively. Figure 6 illustrates changes in the mean trajectory errors across the seven game-sets within an experimental session, compared between groups. A 2 × 2 mixed-model ANOVA revealed a significant interaction effect for group × time [$F(1,39) = 5.26$; $p < 0.05$], comparing the complete reaching movement, but no significant effects for time or group. That means that there was no difference between the groups when combining the two time points, and there was no change in the magnitude of errors when combining the results of the two groups. Pairwise comparisons revealed that the movement error of the study group significantly decreased from 0.036 ± 0.013 m at baseline to 0.029 ± 0.011 m at follow-up ($p < 0.05$), whereas the trajectory errors of the control group did not change from baseline to follow-up. Differences between-group trajectory errors were not significant either at baseline or follow up.

These results, however, were not sustained when adjusting the trajectory errors at baseline as a covariance factor. When comparing performance, there was a significant difference between the two groups in terms of the percentages change from baseline ($p < 0.005$). While the study group trajectories declined by 16.8% from baseline to follow-up, trajectories of the control group increased by 8.5%. Additionally, given our previous experience that indicated a persistent trajectory error of about 0.02 m (Givon-Mayo et al. 2014), these differences were much more rigorous. The measurement of trajectory errors during the first 300 msec of movement revealed no significant differences between groups.

The measurement of the adaptation effect during a game-set was carried out by calculating the trajectory error within a game-set for the initial 300 msec of movement and for the complete movements. Figure 7 illustrates changes in the deviations from the straight trajectory within a training set between the two groups. A 2 × 2 mixed-model ANOVA failed to reveal significant interactions for group × time and for each of the main effects.

4 Discussion

This study investigated whether hand-reaching robotic training with error augmentation force fields, would decrease trajectory errors, compared to training in a null-field environment. Healthy individuals carried out a single-session training of randomly-ordered hand reaching movements for multiple directions within a three dimensional space. We calculated the mean trajectory error for both the complete hand reaching movement and for the early phase of reaching, which consisted of the first 300 msec of each movement. We studied changes in the trajectory errors between game-sets and between movements within a game-set. A separate analysis of these two phases of reaching enabled a discrimination between two different control mechanisms.

During the first 300 msec of movement, the hand trajectory cannot be affected by a correction of movement-errors due to a shortage of time.

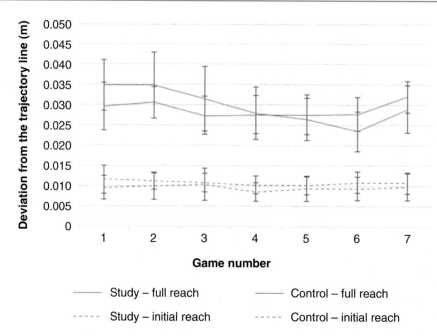

Fig. 6 Mean deviations per game-set of the hand trajectory from the straight line between groups. Dashed lines illustrate the deviations at the first 300 msec of the game set. Complete lines illustrate the mean deviations from the straight line of a complete movement between groups. Error bars indicate the group standard deviations

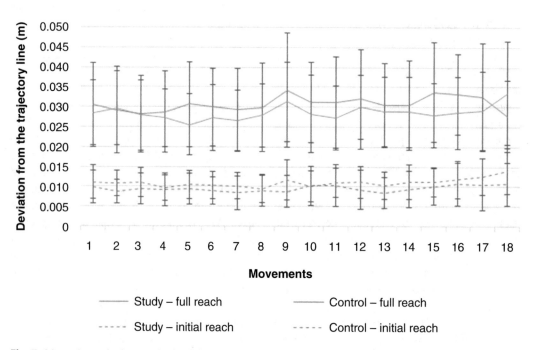

Fig. 7 Mean change in the magnitude of deviation from the straight trajectory line along a game set per group. The group-mean deviation from the straight trajectory line was calculated for each movement during the game-set. Errors were normalized according to the length of the movements. Dashed lines illustrate the deviations at the first 300 msec of a movement. Complete lines illustrate the mean deviations from the straight line of a complete movement. Error bars indicate the group standard deviations

Therefore, trajectory errors at this stage might be attributed to the existing internal model in the brain (Patton et al. 2006b). By contrast, trajectory errors in later stages of movement can be affected by a feedback neural mechanism. The main findings of the present study were those of a significant interaction for group × time, and decreases in the trajectory errors in the study group, while analyzing changes between game-sets. No significant interaction effect nor main effects were found between movements within game-sets.

The lack of a decrease in movement errors between movements within a game-set, in both initial stage of movement and complete movement, is of interest, given the presence of an interaction effect between game sets and a significant improvement seen only in the study group. We had rather expected that the adaptation process would be reflected in a decreased error between movements, especially when training under EA force fields. Therefore, the results suggest that the adaptation mechanism, which involves updating an internal model, did not happen from one movement to another, due assumingly to a variation between consecutive movements. By contrast, force fields applied to the participants of the study group, but not the control group, probably induced a decrease in movement errors from one game-set to another. That might be attributed to enhanced feedback and proprioceptive mechanisms engaged in the study group compared to the control group.

Practicing in null-field environments did not impose any additional perturbation forces to enhance recruitment of control mechanisms to bring the hand toward the target. Thus, the avatar, displayed on the screen, exposed the control group just to a visual feedback of the movement as opposed to an additional proprioceptive feedback experienced by the study group. We assume that these perturbation forces drove the participants to strengthen their control of the movements during the hand-reaching task. Others have suggested that applying EA forces on the hand during reaching tasks increases a signal-to-noise ratio, increases motivation and attention,

and makes the movement errors more perceptible (Patton et al. 2006b). Previous reports have also suggested that encountering perturbation forces induces a Synergistic muscular co-contraction manifested by increasing the perpendicular forces of the hand during movement (Shadmehr and Brashers-Krug 1997).

In terms of trajectory errors between consecutive movements, the present findings are consistent with some previous reports (Huang and Patton 2013), but are discordant with other reports investigating the adaptation process concerning the upper limb (Orban de Xivry and Lefevre 2015; Shadmehr and Moussavi 2000). Still other studies have not reported a change in the magnitude of error from one movement to another (Patton et al. 2006b) or reported conflicting results (Patton et al. 2006a).

Our findings also suggest that the variability between consecutive tasks did not permit generalizability of adaptation. Previous studies have reported the generalizability properties of adaptation across different arm configurations and movement-directions (Donchin et al. 2003; Shadmehr and Moussavi 2000). Others have reported that practicing movements in a random order may impact savings (Huang et al. 2011). Future consideration is whether this kind of training may enhance retention and skill transfer, despite the impact on the short-term task performance (Jonsdottir et al. 2007; Lai et al. 2000; Hanlon 1996).

The experimental tasks, game settings, movement directions, and amplitude and magnitude of forces in the present study were tailored to be used in rehabilitation practice in future studies. Other studies without direct application for rehabilitation may use different measures that may not be applicable for rehabilitation practice. In the present study, duration of hand-reaching training in each experimental session was 20–30 min and consisted of about 100 reaching movements. Based on previous experience with post-stroke EA training (Givon-Mayo et al. 2014), much longer sessions with hundreds of movements are extremely demanding, and in many cases even not possible. Previous studies that investigated adaptation properties consisted of 500–1000

(Donchin et al. 2003), 385 (Orban de Xivry and Lefevre 2015), or 532 (Goodbody and Wolpert 1998) movement trials. Treatment sessions in studies that implemented EA as a rehabilitation technique lasted 60 min (Huang and Patton 2013; Rozario et al. 2009), 3 h or 744 trials (Patton et al. 2006b), or 834 trials (Patton et al. 2006a). Others used 30-min treatment sessions (Molier et al. 2011). Therefore, it is likely that longer sessions in healthy individuals would result in enhanced performance, but would probably not be applicable for post-stroke individuals.

Another aspect of the present treatment setting was to drive long-term improved movement pattern, even at the cost of decreased short-term adaptation effect. Changing task-variables such as direction, amplitude, texture, and load during practice may strike short-term performance, but promotes movement retention and skill transfer (Jonsdottir et al. 2007; Lai et al. 2000; Hanlon 1996). Apparently, a more conservative approach using lower forces with less variability between consecutive movements could be associated with significant adaptation.

Reaching tasks were highly variable within a game-set. Other studies have used limited workspace or two dimensional movements with arm support (Donchin et al. 2003; Sainburg et al. 1999; Conditt et al. 1997) or without arm support (Goodbody and Wolpert 1998). As illustrated in Fig. 3, the direction and length of each consecutive movement was completely random. Therefore, the participant could not anticipate, predict or be prepared for the next movement. Moreover, no two movements within a game-set or even a session were the same. In contrast to the present study, other studies, although employing randomly ordered movement direction, have used fixed directions and distance of targets. The length of the reaching movements of 0.36 m in the present study differed significantly from that reported in previous reports dealing with the simulation of real-life tasks in which it amounted to 0.06 m (Orban de Xivry and Lefevre 2015), 0.10 m (Patton et al. 2006b; Shadmehr and Moussavi 2000), 0.04 m (Krakauer et al. 2005), or 0.14 m (Cesqui et al. 2008). A complete blindness to the next movement, in which each attempt could be considered as the first attempt for a given target location, might increase the demand on the motor system and impact adaptation (Haith and Krakauer 2013; Huang et al. 2011).

As above mentioned, the performance in the first 300 msec of movement relies on the existing internal neural model. Therefore, when practicing movements within 300 msec or less, a significant adaptation effect is anticipated. For instance, practicing a full range of hand reaching movements of 0.36 m during 1.3 s increases the possibility of other learning mechanisms coming into play, which use trial and error to be activated and thus may affect adaptation.

Possible limitations of this study should be taken into consideration. Firstly, movement errors should be analyzed according to the direction of movement and the length of movement. Secondly, a total number of about 100 movements should be doubled to enable clearer improvements between game-sets. In terms of study settings, a mismatch between 3D workspace and 2D screen could expose the participants to ambiguous sensory messages.

5   Conclusions

A hand-reaching robotic training with error augmentation force fields decreases the movement error compared to a training in null-field environments. These differences between the two types of training were manifest across the game-sets, but not within game-sets. Non-significant decreases in the movement errors within a game-set may indicate that the adaptation process failed to generalize between different movement directions and length. That indicates that two mechanisms of learning were simultaneously activated during practice: adaptation and movement-reinforcement procedural learning. Further studies should investigate the application of error augmentation training with longer training protocols.

Acknowledgments We would like to thank Dr. Mario Estevez of the Institute for Neurology and Neurosurgery in Havana, Cuba, for his contributions to statistical and research methodology.

References

Bishop L, Khan M, Martelli D, Quinn L, Stein J, Agrawal S (2017) Exploration of two training paradigms using forced induced weight shifting with the tethered pelvic assist device to reduce asymmetry in individuals after stroke: case reports. Am J Phys Med Rehabil 96:S135–S140

Cesqui B, Macrì G, Dario P, Micera S (2008) Characterization of age-related modifications of upper limb motor control strategies in a new dynamic environment. J Neuroeng Rehabil 5:31

Conditt MA, Gandolfo F, Mussa-Ivaldi FA (1997) The motor system does not learn the dynamics of the arm by rote memorization of past experience. J Neurophysiol 78:554–560

Doig GS, Simpson F (2005) Randomization and allocation concealment: a practical guide for researchers. J Crit Care 20:187–191

Donchin O, Francis JT, Shadmehr R (2003) Quantifying generalization from trial-by-trial behavior of adaptive systems that learn with basic functions: theory and experiments in human motor control. J Neurosci 23:9032–9045

Emken JL, Reinkensmeyer DJ (2005) Robot-enhanced motor learning: accelerating internal model formation during locomotion by transient dynamic amplification. IEEE Trans Neural Syst Rehabil Eng 13:33–39

Givon-Mayo R, Simons E, Ohry A, Karpin H, Israely S, Carmeli E (2014) A preliminary investigation of error enhancement of the velocity component in stroke patients' reaching movements. Int J Ther Rehabil 21:160–168

Goodbody SJ, Wolpert DM (1998) Temporal and amplitude generalization in motor learning. J Neurophysiol 79:1825–1838

Haith AM, Krakauer JW (2013) Model-based and model-free mechanisms of human motor learning. Adv Exp Med Biol 782:1–21

Hanlon RE (1996) Motor learning following unilateral stroke. Arch Phys Med Rehabil 77:811–815

Huang FC, Patton JL (2013) Augmented dynamics and motor exploration as training for stroke. IEEE Trans Biomed Eng 60:838–844

Huang VS, Haith A, Mazzoni P, Krakauer JW (2011) Rethinking motor learning and savings in adaptation paradigms: model-free memory for successful actions combines with internal models. Neuron 70:787–801

Israely S, Carmeli E (2016) Error augmentation as a possible technique for improving upper extremity motor performance after a stroke – a systematic review. Top Stroke Rehabil 23:116–125

Izawa J, Criscimagna-Hemminger SE, Shadmehr R (2012) Cerebellar contributions to reach adaptation and learning sensory consequences of action. J Neurosci 32:4230–4239

Jonsdottir J, Cattaneo D, Regola A, Crippa A, Recalcati M, Rabuffetti M, Ferrarin M, Casiraghi A (2007) Concepts of motor learning applied to a rehabilitation protocol using biofeedback to improve gait in a chronic stroke patient: an AB system study with multiple gait analyses. Neurorehabil Neural Repair 21:190–194

Krakauer JW, Ghez C, Ghilardi MF (2005) Adaptation to visuomotor transformations: consolidation, interference, and forgetting. J Neurosci 25:473–478

Krakauer JW, Carmichael ST, Corbett D, Wittenberg GF (2012) Getting neurorehabilitation right: what can be learned from animal models? Neurorehabil Neural Repair 26:923–931

Lai Q, Shea CH, Wulf G, Wright DL (2000) Optimizing generalized motor program and parameter learning. Res Q Exerc Sport 71:10–24

Lewek MD, Braun CH, Wutzke C, Giuliani C (2017) The role of movement errors in modifying spatiotemporal gait asymmetry post stroke: a randomized controlled trial. Clin Rehabil 1:269215517723056. https://doi.org/10.1177/0269215517723056

Molier BI, Prange GB, Krabben T, Stienen A, van der Kooij H, Buurke JH, Jannink MJ, Hermens HJ (2011) Effect of position feedback during task-oriented upper-limb training after stroke: five-case pilot study. J Rehabil Res Dev 48:1109–1118

O'Brien K, Crowell CR, Schmiedeler J (2017) Error augmentation feedback for lateral weight shifting. Gait Posture 54:178–182

Orban de Xivry JJ, Lefevre P (2015) Formation of model-free motor memories during motor adaptation depends on perturbation schedule. J Neurophysiol 113:2733–2741

Patton JL, Kovic M, Mussa-Ivaldi FA (2006a) Custom-designed haptic training for restoring reaching ability to individuals with poststroke hemiparesis. J Rehabil Res Dev 43:643–656

Patton JL, Stoykov ME, Kovic M, Mussa-Ivaldi FA (2006b) Evaluation of robotic training forces that either enhance or reduce error in chronic hemiparetic stroke survivors. Exp Brain Res 168:368–383

Rozario SV, Housman S, Kovic M, Kenyon RV, Patton JL (2009) Therapist-mediated post-stroke rehabilitation using haptic/graphic error augmentation. Conf Proc IEEE Eng Med Biol Soc 2009:1151–1156

Sainburg RL, Ghez C, Kalakanis D (1999) Intersegmental dynamics are controlled by sequential anticipatory, error correction, and postural mechanisms. J Neurophysiol 81:1045–1056

Shadmehr R, Brashers-Krug T (1997) Functional stages in the formation of human long-term motor memory. J Neurosci 17:409–419

Shadmehr R, Moussavi ZM (2000) Spatial generalization from learning dynamics of reaching movements. J Neurosci 20:7807–7815

Shadmehr R, Mussa-Ivaldi FA (1994) Adaptive representation of dynamics during learning of a motor task. J Neurosci 14(5 Pt 2):3208–3224

Williams CK, Tremblay L, Carnahan H (2016) It pays to go off-track: practicing with error-augmenting haptic feedback facilitates learning of a curve-tracing task. Front Psychol 7:2010

Yen SC, Landry JM, Wu M (2014) Augmented multisensory feedback enhances locomotor adaptation in humans with incomplete spinal cord injury. Hum Mov Sci 35:80–93

Advs Exp. Medicine, Biology - Neuroscience and Respiration (2018) 39: 85–95
DOI 10.1007/5584_2018_150
© Springer International Publishing AG 2018
Published online: 15 Feb 2018

Estimation of Posturographic Trajectory Using *k*-Nearest Neighbors Classifier in Patients with Rheumatoid Arthritis and Osteoarthritis

Beata Sokołowska, Teresa Sadura-Sieklucka, Leszek Czerwosz, Marta Hallay-Suszek, Bogdan Lesyng, and Krystyna Księżopolska-Orłowska

Abstract

Rheumatoid arthritis (RA) and osteoarthritis (OA) are common rheumatic diseases and account for a significant percentage of disability. Posturography is a method that assesses postural stability and quantitatively evaluates postural sways. The objective of this study was to estimate posturographic trajectories applying pattern recognition algorithms. To this end, *k*-nearest neighbors (*k*-NN) classifier was used to differentiate between healthy subjects and patients with OA and RA. The following parameters of trajectories were computed: radius of sways, developed area, total length, and two directional components of sways: length of left-right and forward-backward motions. Posturographic tests were applied with eyes open and closed, and with biofeedback control. We found that in RA, the radius of sways, the trajectory area, and the biofeedback coordination were related to the patients' condition. The trajectory dynamics in OA patients were smaller compared to those in RA patients. The smallest misclassification errors were observed after feature selection in the biofeedback test compared with the eyes open and closed tests. We conclude that the estimation of posturographic trajectory with *k*-NN classifier could be helpful in monitoring the condition of RA patients.

Keywords

Body balance · k-NN classifier · Osteoarthritis · Pattern recognition · Postural stability · Posturography · Rheumatoid arthritis

B. Sokołowska (✉) and L. Czerwosz
Mossakowski Medical Research Centre, Polish Academy of Sciences, Warsaw, Poland
e-mail: beta.sokolowska@gmail.com

T. Sadura-Sieklucka and K. Księżopolska-Orłowska
Rehabilitation Clinic, Professor E. Reicher National Institute Geriatrics Rheumatology and Rehabilitation, Warsaw, Poland

M. Hallay-Suszek
Interdisciplinary Center for Mathematics and Computational Modeling, Warsaw University, Warsaw, Poland

B. Lesyng
Mossakowski Medical Research Centre, Polish Academy of Sciences, Warsaw, Poland

Faculty of Physics, Warsaw University, Warsaw, Poland

1   Introduction

The prevalence of musculoskeletal disorders is 4–5% in the general population (Aletaha et al. 2010; Wong et al. 2010; Sangha 2000). It is estimated that such disorders concern more than one third of the European population. In Poland, these disorders affect every fourth person. The most serious are rheumatoid arthritis (RA) and osteoarthritis (OA) that are highly prevalent in the elderly. A demographic population structure indicates an increasing tendency toward the predominance of the elderly population, with the consequent increase of rheumatic diseases.

RA is a systemic autoimmune disease that affects 0.5–1.0% of adults. A symmetrical inflammatory polyarthritis is the primary clinical manifestation. The arthritis usually begins in small joints of hands and feet, spreading later to larger joints. The inflamed joint lining or synovium extends to, and then erodes, the articular cartilage and bone, causing joint deformity and progressive disability (Gibofsky 2012; Scott et al. 2010). The etiology of RA is still insufficiently known (Westwood et al. 2006; Guidelines (2002). OA, also known as a degenerative joint disorder, affects 10% of men and 18% of women over 60 years of age. OA, is a disease affecting joint cartilage and the underlying subchondral bone. It is characterized by a progressive loss of articular cartilage, appositional new bone formation in the subchondral trabeculae, and a formation of a new cartilage and bone at the joint margins in the form of osteophytes. Pain, stiffness, functional limitation, and diminished quality of life are the primary symptoms associates with OA (Glyn-Jones et al. 2015; Johanson and Hunter 2014).

Posturography is non-invasive technique used to quantitatively estimate the ability of control posture and balance in a broad spectrum of conditions such as physical education, sport training, and in the diagnosis, therapy, or rehabilitation of balance disorders (Paillard and Noé 2015; Arpaia et al. 2014; Visser et al. 2008; Baratto et al. 2002). This technique may also be used for the assessment of RA and OA progression (Sokołowska et al. 2015). The trials conducted on a force plate with eyes open and closed, or under the visual biofeedback coordination are clinically applied in the diagnosis or rehabilitation (Bingham and Calhoun 2015; Czerwosz et al. 2013). The virtual reality technology employed in interactive tasks is a complementary tool in rehabilitation, which effectively supports conventional rehabilitation strategies (Park et al. 2015; Duque et al. 2013; Llorens et al. 2013).

The goal of the present study was to estimate the significance of parameters describing the posturographic trajectories and to evaluate posturographic tests performed for the assessment of body balance stability, with the use of the k-nearest neighbors (k-NN) classifier and statistical pattern recognition algorithms in RA and OA patients.

2   Methods

2.1   Patients and Posturographic Measurements

This clinical study was approved by a local Bioethics Committee of the National Institute of Geriatrics Rheumatology and Rehabilitation in Warsaw, Poland. A patient group consisted of 22 female patients, aged 50–63, with severe multi-joint rheumatic symptoms. The group was subdivided into 11 women with RA and another 11 women with OA. A control group consisted of 11 healthy young women, aged 20–22, displaying no rheumatic signs and symptoms. The evaluation of body balance in the standing position and the measurements of postural sways were carried out by means of a posturographic system (PRO-MED, Legionowo, Poland). The following tests, lasting 32 s each, were applied: with eyes open (EO) and closed (EC), and under conscious visual biofeedback coordination (Fig. 1). The following parameters (features) of recorded trajectories

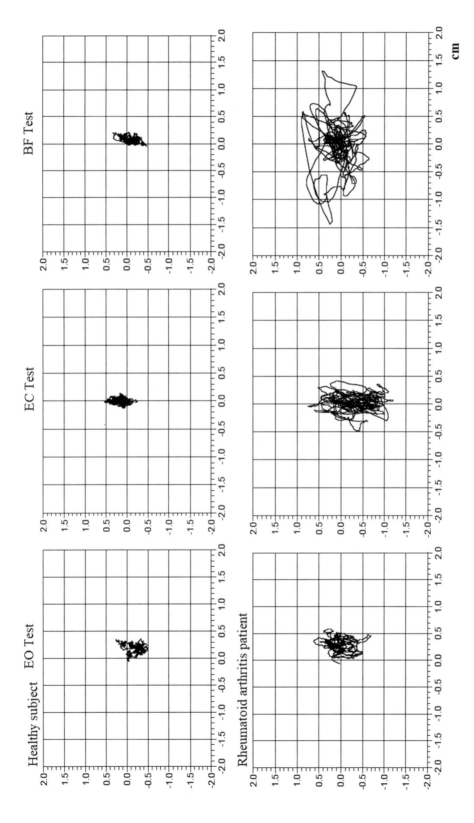

Fig. 1 Examples of posturographic trajectories in the eyes open (EO), eyes closes (EC), and biofeedback (BF) coordination conditions in a healthy subject (upper raw) and rheumatoid arthritis patient (bottom raw)

(posturograms) were computed in all subjects investigated: 1/average radius of sways (R); 2/area of trajectory (A); 3/total length of posturograms (TL); 4/length of left-right motions in a frontal plane (LRL); 5/length of forward-backward motions in a sagittal plane (FBL); and 6/biofeedback coordination, i.e., a percentage of time when the subject's center of pressure (COP) was located within the 10×10 mm visual target square. In the visual BF condition, the subject standing on a force plate could move a marker around a target that might be any object. Tilting the body made the marker moving and mapping an instantaneous COP on the computer screen. An example of configuration with squares as targets, along with COP trajectory, is shown in Fig. 2.

2.2   Pattern Recognition with k-NN Classifier

Pattern recognition is an assignment of labels to objects, e.g. to patients or healthy subjects, which are described by a set of measures called the attributes or features such as clinical parameters.

The pattern recognition methods deal with the objects' classification by a set of features representing the so-called pattern of objects (Duda et al. 2000). Each pattern is represented in terms of n features or measurements and is viewed as a point or vector in the n-dimensional space (Jain 2000). The application of pattern recognition method consists of the following stages: (1) creation of a reference set, consisting of selecting and recording the classes of objects; (2) construction of the decision rule, i.e., a classifier, using the information contained in the reference set, to minimize the misclassification rate, and (3) classification of the objects of uncertain membership using the developed classifier.

Fix and Hodges (1952) have introduced a non-parametric method for the pattern classification that became known as the k-nearest neighbor (k-NN) rule. The k-NN classification is one of the most fundamental and simple classification methods as it is very intuitive and easy to implement for many applications. This classification is based on the measure of distance between objects in the multidimensional feature space. The k-NN rule assigns an object, i.e., a point in the feature

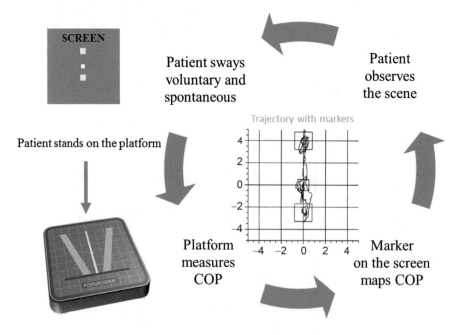

Fig. 2 Schematic diagram of a visual biofeedback (BF) test. COP – center of pressure

space, to the same class as the majority of its *k*-nearest objects in the reference set (Fig. 3).

The classifier quality criterion, depending on the number of the *k*-nearest neighbors, is called the error or misclassification rate (E_r), defined as $E_r = \Delta m/m$, where Δm is the number of misclassified objects, and *m* is a total number of the objects in the reference set. The E_r is calculated for all possible values of *k* using the 'leave one out' method (Duda et al. 2000). This method consists of classifying each of *m* objects from the reference set by the *k*-NN rule derived from the remaining *m-1* objects (Fig. 4). The minimum value of E_r, which is a function of *k*, is selected as a classification and identification quality measure.

The E_r value can be lowered by means of the feature selection (Fig. 5). This procedure rejects redundant features and preserves only the most informative ones. To find the optimum feature subset, it is necessary to review all possible combinations of features and to compute the misclassification error for each of the feature subset reviewed. The result is a combination of features that offer the smallest E_r.

The *k*-NN classifier was used to resolve several experimental and clinical problems in previous studies. That may be exemplified by the recognition of respiratory plasticity in response to exposure to hypoxic stimuli in the animal models (Sokołowska et al. 2003), the evaluation of the effects of mutagenic tests in the *Escherichia coli* model (Maciejewska et al. 2008), the assessment of a progression of amyotrophic lateral sclerosis in patients (Jóźwik et al. 2011; Sokolowska et al. 2009), the identification of differential biomarkers in patients with two forms of Emery-Dreifuss mus-

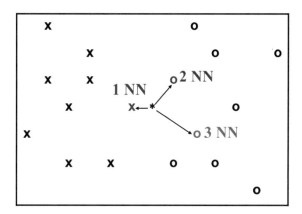

Fig. 3 Illustration of *k*-NN rule. 'x' – points belonging Class I and 'o' – points belonging to Class II. The symbols *1*-NN, *2*-NN, and *3*-NN denote respectively: the first, the second, and the third nearest neighbor of a new classified point (denoted by „*"). According to the *3*-NN rule, the point „*" is assigned to Class II since two, out of its three, nearest neighbors come from Class II ("x" – points from Class I, "o" – points belonging to Class II)

1		2		3			4	5	6			7	8	9			10
x		x		x			x	o	x			o	o	o			o

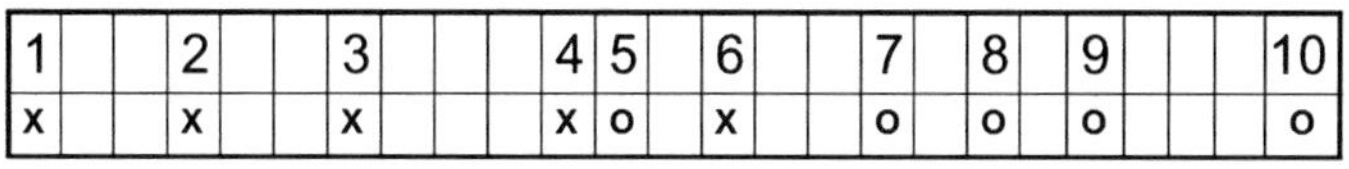

Fig. 4 Illustration of the 'leave one out' method. 'x' - points belonging to class I and 'o' - points belonging to class II. The *1*-NN rule misclassifies three points: 4, 5, and 6, whereas the *3*-NN rule misclassifies only two points: 5 and 6

Fig. 5 Model of the statistical pattern recognition method employed in the study

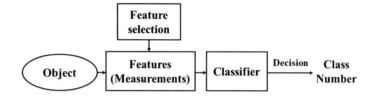

Table 1 Summary description of classes, features, and tests

Class	Feature	Test (conditions)
I – Healthy subjects	1 – R	EO test
	2 – A	
II – OA patients	3 – TL	EC test
	4 – LRL	
III – RA patients	5 – FBL	BF test
	6 – BF	

OA osteoarthritis, *RA* rheumatoid arthritis, *R* average radius of sways, *A* developed area of trajectory, *TL* total length of trajectory. Two directional components of sways: *LRL* length of left-to-right frontal plane movements and FBL – forward-backward length of movements in sagittal plane, *EO* eyes open, *EC* eyes closed, *BF* biofeedback coordination parameter

cular dystrophy (Sokołowska et al. 2014), and the differential diagnosis of patients with normal pressure hydrocephalus and brain atrophy (Czerwosz et al. 2013).

In the present study, the recognition (differentiation) task of patients with RA and OA according to their posturographic trajectory was realized by applying three different tests (conditions) on a force plate: EO, EC, and BF. The three classes (healthy subjects, RA patients, and OR patients) and 5 or 6 features were defined (Table 1). In the first step, we constructed a classifier based on the *k*-NN rule and then we calculated the E_r using the 'leave one out' method. The three data sets were applied for the training task: EO, EC, and BF.

The structure of the parallel *k*-NN classifier for the three classes is presented in Fig. 6. The classifier consists of three two-decision *k*-NN classifiers. The final decision is formed by voting for the component classifiers. The class that gathers the greatest number of votes is selected. The analysis was performed without and with feature selection.

3   Results

The E_r for class differentiation (healthy *vs.* patient groups) was calculated for each single feature (posturographic parameter) and then for a set of all features in each posturographic test condition, i.e., EO, EC, and BF. The E_r values were significantly smaller after feature selection compared with the values without selection.

3.1   Eyes Open (EO) Test

The lowest E_r was observed for Feature 1, i.e., R – radius of the trajectory, in the EO test. This feature enables the differentiation between the healthy controls (Class I) and both patient Classes II and III with the E_r of 0.136. The misclassification error is significantly larger when all features are analyzed, amounting to 0.227. In this test, higher E_r values were observed for the class recognition after feature selection than those in other tests (Table 2).

3.2   Eyes Closed (EC) Test

The smallest E_r of 0.046 was observed for Feature 2, i.e., A – developed area of trajectory, in the EC test. This feature enables the differentiation between Classes I and III, i.e., healthy controls and RA patients. The differentiation between Classes I and II, i.e., healthy controls and OA patients, for the feature set {2,3,4} provided a two-fold greater E_r of 0.091, but both results were significantly better than those in the EO test. It remained still difficult to differentiate between Classes II and III, although Feature 1, i.e., R – radius of the trajectory, provided a lower E_r of 0.182, compared with the 0.318 in the

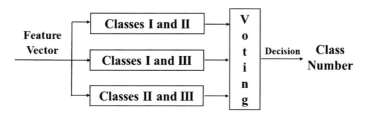

Fig. 6 Structure of the *k*-NN classifier for three classes (I, II, and III). It consists of three two-decision *k*- NN classifiers (i.e., I/II, I/III, and II/III) and the final decision is achieved by voting

Table 2 Misclassification errors (E_r) for the class differentiation (recognition) in the eyes open (EO) test

Feature – EO test	E_r			
	(I and II)	(I and III)	(II and III)	(I, II, and III)
1 – R	0.136	0.136	0.318	0.303
2 – A	0.182	0.136	0.455	0.394
3 – TL	0.318	0.227	0.591	0.546
4 – LTR	0.364	0.318	0.500	0.576
5 – FBL	0.318	0.273	0.455	0.515
{1, 2, 3, 4, 5}	0.227	0.227	0.455	0.455
After feature selection	0.136	0.136	0.318	0.303
{Set of selected features}	{1}	{1}	{1}	{1}

Features and classes are in accordance with those displayed in Table 1

EO test (Table 2). Also, E_r decreased approximately by half for all classes when the patients performed the task with eyes closed; a decrease was from 0.303 to 0.152. After the feature selection, all E_r decreased significantly (Table 3).

3.3 Biofeedback (BF) Coordination Test

A straightforward differentiation between Classes I and III, i.e., healthy controls and RA patients, was observed in the BF test. Every feature, excluding Feature 5 ($E_r = 0.182$), provided the E_r smaller than 5%, and even Feature 4 or 6 provided a perfect differentiation ($E_r = 0.0001$). The feature set {1,3,6} provided a lower E_r for the differentiation of both patient classes, i.e., Classes II and III, compared with the E_r in the other tests. It remained still difficult to differentiate between the two patient classes, although the feature selection improved the differentiation quality (Table 4).

4 Discussion

Two most common rheumatic diseases, RA and OA lead to a destruction of the motor system, cause pain, weakness, as well as damage the ligaments, muscles, bones, and articular cartilage. In the current study, we attempted to differentiate patients with RA and OA using a posturographic approach with new analytical algorithms. The trajectory variables were estimated in the tests applying statistical pattern recognition algorithms, with the *k*-NN classifier. The analysis demonstrates that the radius (R) differentiated all the study groups, i.e., control subjects, and RA and OA patients, in the eyes closed condition. A feature selection confirmed the importance of the same single feature (R). The feature enabled differentiation of patients, regardless of the type of rheumatic disease, from healthy individuals. The estimation of the R posturographic trajectory may be helpful in monitoring patients, for instance, before and after therapy or rehabilitation. Of note, the greatest misclassification error was observed in the eyes open condition. The

Table 3 Misclassification errors (E_r) for the class differentiation (recognition) in the eyes closed (EC) test

Feature – EC test	E_r			
	(I and II)	(I and III)	(II and III)	(I, II, and III)
1 – R	0.136	0.227	0.182	0.273
2 – A	0.182	0.046	0.364	0.273
3 – TL	0.182	0.227	0.445	0.394
4 – LTR	0.227	0.227	0.364	0.424
5 – FBL	0.182	0.318	0.409	0.455
{1, 2, 3, 4, 5}	0.182	0.273	0.364	0.364
After feature selection	0.091	0.048	0.182	0.152
{Set of selected features}	{2, 3, 4}	{2}	{1}	{1, 2, 3, 4}

Features and classes are in accordance with those displayed in Table 1

Table 4 Misclassification errors (E_r) for the class differentiation (recognition) in the biofeedback (BF) coordination test

Features – BF test	E_r			
	(I and II)	(I and III)	(II and III)	(I, II, an d III)
1 – R	0.273	0.046	0.273	0.364
2 – A	0.227	0.046	0.182	0.274
3 – TL	0.136	0.046	0.227	0.242
4 – LTR	0.182	0.0001	0.227	0.273
5 – FBL	0.227	0.182	0.273	0.333
6 – BF	0.134	0.0001	0.318	0.303
{1, 2, 3, 4, 5, 6}	0.227	0.0001	0.227	0.303
After feature selection	0.134	0.0001	0.136	0.182
{Set of selected features}	{6}	{4} or {6}	{1, 3, 6}	{1, 3, 4 ,6}

Features and classes are in accordance with those displayed in Table 1

R feature also maintained its effectiveness in differentiating RA from OA patients in the eyes closed condition, with the misclassification error reduced by 50%. In addition, features R and A appeared effective in differentiating healthy subjects from both patient groups. The BF test had the greatest efficacy in differentiating the groups examined. Also, BF coordination parameter appeared effective in the differentiation procedure of the two rheumatic pathologies.

Other studies have reportedly demonstrated the potential usefulness of posturography in clinical practice of rheumatic patients (Negahban et al. 2016; Zhang et al. 2015; Park et al. 2013; Chaudhry et al. 2011; Kim et al. 2011). Several measurement algorithms for the assessment of trajectories and for data analysis in posturography have been reported (Błaszczyk 2016; Cretual 2015; Piirtola and Era 2006; Baratto et al. 2002). Thus, there are a variety of approaches to

these measurements and their interpretations concerning their suitability in practice (Answer et al. 2015; Błaszczyk et al. 2014; Brenton-Rule et al. 2014). The studies are usually carried out using computer recording systems, with online or offline analysis. Such systems are often adapted to the disease conditions or specific features of objects being examined in both experimental models and clinic trials. Hen et al. (2000) have studied the influence on the standing balance of double tasks in patients with RA, accompanied by severe knee joint impairment. Those authors estimated postural sways during quiet standing with eyes open/closed and while performing a secondary attention-demanding arithmetic task. Differences in the velocity-related parameters between patients and controls were analyzed by a multivariate analysis of variance. Patients with RA swayed significantly stronger than control subjects. A superimposed effect of the arithmetic tasks was negligible and similar for both groups of subjects. The authors conclude that RA causes a substantial basic postural instability. In turn, Kim et al. (2011) have studied balance control in patients with mild OA and with moderate-to-severe OA, in comparison with age-matched controls. The authors defined eight different posturographic tests, two with eyes open and six with eyes closed, and chose for the analysis of postural variables the indices of stability, Fourier, weight distribution, and synchronization. Classical statistics showed that patients with moderate-to-severe OA exhibited a significantly higher stability in all positions (tests) than patients with mild OA, which correlated with a greater decrease in muscle strength, proprioception, and increased pain, all contributing to postural instability in the milder form of OA. Park et al. (2013) have assessed clinical factors and calculated variables related to the standing balance in females with OA, with eyes open/closed for 30 s. The mean speed of COP in the anteroposterior and mediolateral directions was computed. A univariate regression analysis was carried out to assess effects of age, pain, knee alignment and the severity of radiographic changes on posturographic parameters. The findings suggest that a greater balance impairment is mainly associated with advanced age. Zhang et al. (2015) have demonstrated that balance stability in patients with OA knee deformity is different during day times. The authors suggest that the altered postural performance in the morning could have to do with the joint pain. Diurnal variations should thus be taken into account in the daily management of OA patients. Negahban et al. (2016) have employed non-linear statistical methods to investigate differences in the complexity and variability of sway dynamics between OA patients and healthy subjects under four different conditions of postural (single) plus cognitive (dual) tasks, with EO or EC. The analysis demonstrates less complexity and more variability of postural sways in OA patients compared with healthy subjects. Moreover, a non-linear behavior of both groups showed a decreased complexity and increased variability under challenging sensory conditions, while increased complexity and decreased variability were observed during dual compared with single-task conditions. Sokołowska et al. (2015) have analyzed several posturographic trajectory parameters akin to those assessed in the present study (R, A, LRL, FBL, TL, and BF) in the EO, EC, and BF tests in patients with OA and RA in comparison with healthy subjects. The results of an extended classical statistical analysis accounting for the receiver operating characteristic (ROC) curves show that the patients exhibited significantly greater postural sways with both EO and EC, compared with healthy subjects. Postural sways were also greater in RA than OA patients. The BF test appeared superior to the others, with the sensitivity and specificity values of about 0.77 for both RA and OA patients.

In synopsis, the radius, developed area of trajectory, and biofeedback coordination parameters were related to the patient status in rheumatic diseases during posturographic examinations. Visual tasks under the biofeedback control, which refer to the balance stability, appear the valuable procedures in clinical practice of musculoskeletal pathologies. We conclude that posturography is a non-invasive, simple and effective method to detect disorders of the motor system. This technique, combined with the pattern recognition algorithms, enables a quantitative evaluation of the balance control. Thus, it may be a worthwhile tool for clinicians and physiotherapists in dealing with rheumatic diseases.

Acknowledgments We thank Dr. A. Jóźwik for making his *k*-NN software available for this study and Dr. F. Rakowski for valuable remarks concerning the posturographic trajectories. The work was supported by grant MMRC PAS and the Faculty of Physics of Warsaw University (grant BST-1733000/bf task 34).

Conflicts of Interest The authors declare no conflicts of interest in relation to this article.

References

Aletaha D, Neogi T, Silman AJ, Funovis J, Felson DT, Bingham COIII et al (2010) Rheumatoid arthritis classification criteria: an American College of Rheumatology/European League Against Rheumatism collaborative initiative. Ann Rheum Dis 69:1580–1588

Answer S, Alghadir A, Brismee JM (2015) Effect of home exercise program in patients with knee osteoarthritis: a systematic review and meta-analysis. J Geriatr Phys Ther 39(1):38–48

Arpaia P, Cimmino P, De Matteis E, D'Addio G (2014) A low-cost force sensor-based posturographic plate for home care telerehabilitation exergaming. Measurement 51:400–410

Baratto L, Morasso PG, Re C, Spada G (2002) A new look at posturographic analysis in the clinical context: sway-density vs. other parametrization techniques. Motor Contr 6:246–270

Bingham PM, Calhoun B (2015) Digital posturography games correlate with gross motor function in children with cerebral palsy. Games Health J 4(2):1–4

Błaszczyk JW (2016) The use of force-plate posturography in the assessment of postural instability. Gait Posture 44:1–6

Błaszczyk JW, Beck M, Sadowska D (2014) Assessment of postural stability in young healthy subjects based on directional features of posturographic data: vision and gender effects. Acta Neurobiol Exp 74(4):433–442

Brenton-Rule A, D'Almeida S, Basset S, Carroll M, Dalbeth N, Rome K (2014) The effects of sandals on postural stability in patients with rheumatoid arthritis: an exploratory study. Clin Biomech 29:350–353

Chaudhry H, Bukiet B, Ji Z, Findley T (2011) Measurement of balance in computer posturography: comparison of methods – a brief review. J Bodyw Mov Ther 15(1):82–91

Cretual A (2015) Which biomechanical models are currently used in standing posture analysis? Neurophysiol Clin 45:285–295

Czerwosz L, Szczepek E, Sokołowska B, Jurkiewicz J, Czernicki Z (2013) Posturography in differential diagnosis of normal pressure hydrocephalus and brain atrophy. Adv Exp Med Biol 755:311–324

Duda OR, Hart PE, Stork DG (2000) Pattern classifcation. Wiley Interscience, New York

Duque G, Boersma D, Loza-Diaz G, Hassan S, Suarez H, Geisinger D, Suriyaarachchi P, Sharma A, Demontiero O (2013) Effects of balance training using a virtual-reality system in older fallers. Clin Interv Aging 8:257–263

Fix E, Hodges JL (1952) Discriminatory analysis: non-parametric discrimination small sample performance. Project 21-49-004, Report Number 11, USAF School of Aviation Medicine, Randolph Field, Texas

Gibofsky A (2012) Overview of epidemiology, pathophysiology, and diagnosis of rheumatoid arthritis. Am J Manag Care 18:S295–S302

Glyn-Jones S, Palmera AJ, Agricola R, Price AJ, Vincent TL, Carr AJ (2015) Osteoarthritis. Lancet 386(9991):376–387

Guidelines (2002) Management of rheumatoid arthritis -2002 update. Arthritis Rheum 46(2):328–346

Hen SS, Geurts AC, van't Pad BP, Laan RF, Mulder T (2000) Postural control in rheumatoid arthritis patients scheduled for total knee arthroplasty. Arch Phys Med Rehabil 81:1489–1493

Jain AK (2000) Statistical pattern recognition: a review. IEEE Trans Pattern Anal Mach Intell 22(1):4–37

Johanson VL, Hunter DJ (2014) The epidemiology of osteoarthritis. Best Pract Res Clin Rheumatol 28:5–15

Jóźwik A, Sokołowska B, Niebroj-Dobosz I, Janik P, Kwieciński H (2011) Extraction of biomedical traits for patients with amyotrophic lateral sclerosis using parallel and hierarchical classifiers. Int J Biometrics 3(1):85–94

Kim HS, Yun DH, Yoo SD, Kim DH, Jeong YS, Yun JS, Hwang DG, Jung PK, Choi SH (2011) Balance control and knee osteoarthritis severity. Ann Rehabil Med 35:701–709

Llorens R, Colomer-Font C, Alcaniz M, Noe-Sebastian E (2013) BioTrak virtual reality system: effectiveness and satisfaction analysis for balance rehabilitation in patients with brain injury. Neurologia 28(5):268–275

Maciejewska A, Jóźwik A, Kuśmierek JT, Sokołowska B (2008) Application of the k-NN classifier for mutagenesis test. Recognition of the wild type and defective in DNA repair bacterial strains on the basis of adaptive response to alkylating agents. Biocybern Biomed Eng 28(3):45–50

Negahban H, Sanjari MA, Karimi M, Parnianpour M (2016) Complexity and variability of the center of pressure time series during quiet standing in patients with knee osteoarthritis. Clin Biomech 32:280–285

Paillard T, Noé F (2015) Techniques and methods for testing the postural function in healthy and pathological subjects. Biomed Res Int 2015:891390

Park HJ, Ko S, Hong HM, Ok E, Lee JI (2013) Factors related to standing balance in patients with knee osteoarthritis. Ann Rehabil Med 37(3):373–378

Park EC, Kim SG, Lee CW (2015) The effects of virtual reality game exercise on balance and gait of the elderly. J Phys Ther Sci 27:1157–1159

Piirtola M, Era P (2006) Force platform measurements as predictors of falls among older people – a review. Gerontology 52(1):1–16

Sangha O (2000) Epidemiology of rheumatic diseases. Rheumatology 39(Suppl 2):3–12

Scott DL, Wolfe F, Huizinga TW (2010) Rheumatoid arthritis. Lancet 376(9746):1094–1098

Sokołowska B, Jóźwik A, Pokorski M (2003) A fuzzy-classifier system to distinguish respiratory patterns evolving after diaphragm paralysis in the cat. Jpn J Physiol 53(4):301–307

Sokolowska B, Jozwik A, Niebroj-Dobosz I, Janik P, Kwiecinski H (2009) Evaluation of matrix metalloproteinases in serum of patients with amyotrophic lateral sclerosis with pattern recognition methods. J Physiol Pharmacol 60(Suppl 5):117–120

Sokołowska B, Jóźwik A, Niebroj-Dobosz I, Hausmanowa-Petrusewicz I (2014) A pattern recognition approach to Emery-Dreifuss muscular dystrophy (EDMD) study. MIT J 23:165–171

Sokołowska B, Czerwosz L, Hallay-Suszek M, Sadura-Sieklucka T, Księżopolska-Orłowska K (2015) Posturography in patients with rheumatoid arthritis and osteoarthritis. Adv Exp Med Biol 2:63–70

Visser JE, Carpenter MG, van de Kooij H, Bloem BR (2008) The clinical utility of posturography. Clin Neurophysiol 119(11):2424–2436

Westwood OM, Nelson PN, Hay FC (2006) Rheumatoid factors: what's new? Rheumatology (Oxford) 45 (4):379–385

Wong R, Davis AM, Badley E, Grewal R, Mohammed M (2010) Prevalence of arthritis and rheumatic diseases around the world. A growing burden and implications for health care needs. Arthritis community research and evaluation unit. http://www.acreu.ca/moca. Accessed 20 Oct 2017

Zhang Z, Lion A, Chary-Valckenaere I, Loeuille D, Rat A-K, Paysant J, Perrin PP (2015) Diurnal variation on balance control in patients with symptomatic knee osteoarthritis. Arch Gerontol Geriatr 61(1):109–114

Advs Exp. Medicine, Biology - Neuroscience and Respiration (2018) 39: 97–109
DOI 10.1007/5584_2018_153
© Springer International Publishing AG 2018
Published online: 13 Feb 2018

Effects of Manual Somatic Stimulation on the Autonomic Nervous System and Posture

Giovanni Barassi, Rosa Grazia Bellomo, Camillo Di Giulio, Giuseppe Giannuzzo, Giuseppe Irace, Claudia Barbato, and Raoul Saggini

Abstract

Low back pain frequently involves a multifactorial etiology and requires medical attention. The aim of the study was to assess the associations among pain, posture, and autonomic nervous system function in patients with low back pain, using neuromuscular manual therapy *versus* a generic peripheral manual stimulation (back massage therapy). Twenty young patients with low back pain were enrolled into the study. The patients were randomly divided into two groups: treated with neuromuscular manual therapy performed after a specific structural evaluation and treated with back massage therapy. Both groups performed eight sessions of 30 min each, once a week for two months. There were three main time points of the assessment: during the first, the fourth, and the last eighth session. In each of these three sessions, data were collected before onset of session (baseline), 5 min from onset, at end of session, and 5 min after the end. All patients were subjected to stabilometric evaluation and were assessed on a visual analogue scale to quantify postural and pain changes.

Tabletop capnography and pulse oximetry were used to monitor autonomic changes. The findings were that the improvement in posture and pain reduction were appreciably better in patients subjected to neuromuscular manual therapy than in those subjected to back massage therapy, with a comparable autonomic response in both groups. In conclusion, the study demonstrates that posture modification was significantly more advantageous in patient treated with neuromuscular manual therapy.

Keywords

Autonomic nervous system · Low back · Manual therapy · Massage therapy · Neuromuscular dysfunction · Pain · Posture · Stabilometry · Structural evaluation · Trigger point

1 Introduction

Low back pain remains a condition with a high incidence and prevalence. Low back pain is

G. Barassi (✉), G. Giannuzzo, G. Irace, C. Barbato, and R. Saggini
Department of Medical Oral and Biotechnological Science, "Gabriele d'Annunzio" University, Chieti-Pescara, Italy
e-mail: coordftgb@unich.it

R. G. Bellomo
'Carlo Bo'-University, Urbino, Italy

C. Di Giulio
Department of Neuroscience and Imaging, "Gabriele D'Annunzio" University, Chieti-Pescara, Italy

primarily responsible for more than 20 million ambulatory medical care visits and $100 billion in annual cost in the United States alone (Licciardone 2008; Katz 2006). Following a first episode, pain typically improves substantially, but it does not resolve completely over the following 4–6 weeks. In most patients, pain and associated disability persist for months. However, only a small proportion remain severely disabled. For those whose pain does resolve completely, recurrence during the following 12 months is not uncommon (Koes et al. 2006; Pengel et al. 2003).

Non-specific low back pain is defined as pain that cannot be attributed to a recognizable pathology, e.g., infection, tumor, osteoporosis, fracture, structural deformity, inflammatory disorder, radicular syndrome, or cauda equina syndrome (Balagué et al. 2012). Patients with non-specific low back pain may have somatic dysfunction as the cause or contributing factor of pain. The diagnosis of somatic dysfunction encompasses a history of symptoms and physical examination, including a structural examination that provides evidence of asymmetrical anatomic landmarks, restriction or altered range of joint motion, and palpatory abnormalities of soft tissues. Treatment for somatic dysfunction is initiated after other potential causes of low back pain are ruled out or considered improbable, such as vertebral fracture, vertebral joint dislocation, muscle tears or lacerations, spinal or vertebral joint ligament rupture, inflammation of intervertebral disks, spinal zygapophyseal facets joints, muscles, or fascia; skin lacerations, sacroiliitis, ankylosing spondylitis, pathological mass arising in or from the low back structures, or organic (visceral) disease causing pain in the back or low back muscle spasms (American Osteopathic Association 2010).

The literature describes various disruptions in the pattern of recruitment and co-contraction within and between different muscle synergies in case of low back pain. There have also been reports that compensatory substitution of global system muscles occurs in the presence of local muscle dysfunction. This compensation appears to be a neural control system's attempt to maintain stability requirements of the spine in the presence of local muscle dysfunction. There is evidence that the presence of chronic low back pain often results in a general loss of function and

de-conditioning, and in changes to the neural control system, affecting timing of co-contraction, balance, and reflex and righting responses (O'Sullivan et al. 1997). Such disruptions to the neuro-muscular system leave the lumbar spine potentially vulnerable to instability, particularly within the neutral zone (Cholewicki and McGill 1996).

Non-specific low back pain is more than epiphenomenon and represents a continuous source of afferent barrage. The autonomic nervous system is involved in the control of heart, glands, and smooth muscles and it plays a major role in the regulation of unconsciously performed functions. This system works along with somatic nervous system, as motor fibres make up the bulk of the autonomic system. Somatic dysfunction is defined as impaired or altered function of body framework components, such as skeletal, arthrodial, and myofascial structures, and related vascular, lymphatic, and neural elements (Cervero and Connell 1984). It is proposed to be a reversible, functional disturbance that predisposes the body to disease, in which a myofascial manipulation constitutes an effective treatment (Schleip 2003). The term somatic dysfunction can be used broadly to denote dysfunction of a group of tissues or a region, or more specifically to denote dysfunction of a single articulation. Somatic dysfunction is not synonymous with spinal pain, as the palpable signs of dysfunction may be detected in symptomatic and asymptomatic individuals (Fryer et al. 2004; Saggini and Ridi 2002). The presence of somatic dysfunction in asymptomatic individuals creates biomechanical and neurological consequences that predispose to pain and other health complaints (Patterson and Wurster 2011; Travell and Simons 1992).

The objective of this study was to evaluate and validate a specific somatic stimulation treatment in patients with non-specific low back pain caused by an underlying somatic dysfunction. The evaluation outcomes consisted of autonomic nervous system responses, postural changes, and pain perception. These outcomes were compared with the effects of a simple generic peripheral stimulation (spine massage therapy). We used a specific structural evaluation method that provides evidence of asymmetrical anatomic landmarks, restriction or altered range of joint motion, and the palpatory abnormalities of soft tissues.

2   Methods

2.1   Patients and Instrumentation

The study was conducted in accordance with the requirements set by the Ethics Committee 'Comitato Etico per la Ricerca Biomedica' of "G. d'Annunzio" University in Chieti-Pescara, Italy, and with the Declaration of Helsinki for Human Research. The patients were informed about the study procedures and gave informed written consent. The study data were stored at the Center of Physical Medicine and Rehabilitation of Chieti University. There were 20 non-specific low back pain patients aged 22–29 (average age 25 years) enrolled into the study. They were randomly divided into two groups of 10 patients each: Group A, treated with neuromuscular manual therapy performed after a specific structural evaluation, and Group B, treated with back massage therapy, used as a reference group for comparison with Group A. All patients were subjected to stabilometric evaluation and postural and pain changes were quantified on a visual analogue scale. Tabletop capnography and pulse oximetry (EtCO$_2$; SpO$_2$) were used to monitor autonomic variables before and after the sessions above outlined. A latency of 5 min from onset of monitoring was employed to enable the adjustment of variables if required.

The visual analog scale for pain assessment is a psychometric instrument useful to quantify subjective characteristics or attitudes that cannot be directly measured, e.g., perceived pain intensity, on a numeric scale from 0 to 10 (Price et al. 1983). Stabilometry is an objective assessment of body sways during quiet standing in the absence of any voluntary movements or external perturbations. The method enables the collection of information on the steady-state functioning of the postural control system and its ability to stabilize the body against gravity. This evaluation is performed using specific computerized boards that record body's

postural adjustments with high grade of sensitivity (Gori and Firenzuoli 2005; Kapteyn et al. 1983). Capnography monitors inhaled and exhaled content of CO$_2$, and thus, indirectly, partial pressure of CO$_2$ in the arterial blood. The difference between the arterial and expired CO$_2$ is minimal in healthy people, while an increase or decrease of this difference speaks for a systemic or localized health problem. We used an infra-red capnograph, where the absorption of infra-red light by CO$_2$ is proportional to the content of CO$_2$. Peripheral oxygen saturation (SpO$_2$) was assessed with an oximeter (Capnografo con Saturimetro Lifesense; NONIN Medical, RAM Apparecchi Medicali, Genova, Italy).

2.2   Study Protocol

The patients of both groups were subjected to eight sessions of treatment, lasting for 30 min each, one session per week for 2 months. There were three main time points of the assessment: during the first session (T0), the fourth session (T1), and the protocol ending eighth session (T2). In each of the three assessment sessions, data were collected before onset of session (baseline), 5 min from onset, at end of session, and 5 min after the end, which was marked as X1, X2, X3, and X4, respectively.

Group A patients were treated with specific manual neuromuscular therapy focusing on the areas of somatic dysfunction detected during structural evaluation:

– sternocleidomastoid and levator scapula muscle – five patients;
– foot plantar region and quadratus lumborum muscle – two patients;
– trapezius, sternocleidomastoid, levator scapula, and quadratus lumborum muscle – three patients.

Therapy involved the following maneuvers:

– relaxation warming – quick and gentle heating to create a hyperemic tissue condition, like effleurage and petrissage; skin rolling and circular friction to prepare the patient for direct maneuvers;

- direct maneuvers – compression using an hook grip or pincer with a pressure of about 3 kg, and a gradual decrease of pressure without abruptly stopping the contact.

Muscles selected for treatment were manipulated using the following techniques:

- upper trapezius muscle – treated in the supine position, followed by prone position with the scapula and shoulder in a neutral position. A direct compression was performed on the upper and middle trapezius.
- upper part of the upper trapezius muscle – palpation following the direction of muscle fibers, starting from the acromion-clavicular joint, continuing along the muscle to the distal insertion. Direct compression was performed along the belly of the muscle.
- middle part of the upper trapezius muscle – palpation following the direction of muscle fibers, starting at the level of the spine, scapula and the acromion, with pressure exerted downwards toward the upper area of the scapula and upper chest, continuing the palpation over the entire length of fibers. Direct compression was performed along the belly of the muscle.
- medium trapezius – treated in the lateral decubitus position, anteposition and abduction of the shoulder; exercises were performed while sitting, elevating the upper limbs to the forehead, then adducing the shoulder horizontally and stabilizing the distal humerus with one hand.
- levator scapula – compression along the direction of muscle fibers, starting at the medial border of the scapula above the spine of the scapula; compression was carried out through the trapezius in the direction of upper chest, posterior to the insertion.
- quadratus lumborum muscle – treated in the lateral decubitus position, with the lumbar spine in lateral flexion, the hip in extension, and the shoulder in abduction; performed through direct palpation along the lateral border of the muscle, starting from the area between the twelfth rib and the iliac crest. The pressure was directed medially through the muscle in the area

of the third side portion of the lumbar spine. The compression was performed in medial, lateral, and caudal directions, and to the junction of the fifth lumbar vertebra and the posterior iliac crest of the pelvic girdle. The treatment continued by exerting pressure in medial, lateral, and cranial directions to the junction of the first lumbar vertebra with the twelfth rib.

- sternocleidomastoid muscle – treated in the supine position, with the head slightly down and in a contralateral rotation with regard the side being treated. A pincer compression was done in the direction of muscle fibers in the proximal-distal direction.
- plantar foot region – specific treatment of the most important muscles of the plantar area.
- abductor halluces – compressions were exerted in the supine position, with flexed knees, directly on the central portion of the muscle just below the first metatarsal region, starting at the back of the heel. A direct pressure was applied on the first metatarsal and continued medially to the distal insertion.
- brevis digitorum flexor – compression was done in three lines identified in approximately three/fifth of the plantar region, which was completed compression in the distal-proximal direction, perpendicular to the surface of the muscle on the plantar fascia.
- halluces brevis flexor – compression was done along the two lines on the plantar surface of the foot, which was applied in the distal proximal direction, perpendicular to the examined surface and deep on the plantar fascia, continuing up to the insertion.

2.3 Statistical Analysis

Presented data are means $\pm$ SD. The significance of differences between T0, T1, and T2 time points was assessed with one-way ANOVA, while differences between the mean values of a single treatment session were assessed with one-way ANOVA for repeated measures. A p-value < 0.05 defined statistically significant differences.

3   Results

The variables measured, such as heart rate, breaths *per* minute, or oxygen saturation, generally improved significantly in both groups of patients. In Group A, treated with neuromuscular manual therapy, the mean heart rate at T0 was 73.5 $\pm$ 1.0 beats/min, breathing rate was 18.4 $\pm$ 1.8 breaths/min, and SpO_2 was 96.5 $\pm$ 0.5%. These values changed to 66.1 $\pm$ 0.9 beats/min, 13.8 $\pm$ 1.1 breaths/min, and 98.0 $\pm$ 0.4%, respectively at TI ($p < 0.05$). In group B, treated with back massage therapy, the mean heart rate at T0 was to 69.5 $\pm$ 1.0 beats/min, breathing rate was 18.1 $\pm$ 1.7 breaths/min, and SpO_2 was 97.3 $\pm$ 0.6%. These values changed to 56.8 $\pm$ 0.9 beats/min, 11.3 $\pm$ 1.0 breaths/min, and 98.9 $\pm$ 0.5, respectively, at T1 ($p < 0.05$).

Heart rate, on average, showed a significant decrease from T0 to T1 sessions in both patient groups (Fig. 1). A decrease in heart rate also was evident when it was evaluated at the four time marks of each main session, i.e., onset (baseline), 5 min from onset, end, and 5 min after session end, marked as X1, X2, X3, and X4, respectively (Fig. 2). Likewise, breathing rate showed a significant decrease from T0 to T1 sessions in both patient groups (Fig. 3) as well as in the subanalysis of the four time marks of each main session (Fig. 4).

Likewise, oxygen saturation showed a significant decrease from T0 to T1 sessions in both patient groups (Fig. 5) and when it was evaluated at the four time marks of each main session, i.e., onset (baseline), 5 min from onset, end, and 5 min after the session end, marked as X1, X2, X3, and X4, respectively (Fig. 6).

With respect to stabilometry, significant changes were observed only in Group A, in which the patients were subjected to the manual therapy focusing on a specific myofascial dysfunction. In this group, the ellipse surface in the evaluation made with open eyes on the platform was equal to 173.5 $\pm$ 14.5 mm^2 at T0; this value was reduced to 161.9 $\pm$ 8.3 mm^2 at T1 ($p < 0.05$). In group B, the value for the ellipse surface with open eyes tended also to be reduced; an insignificant change from 174.5 $\pm$ 17.3 mm^2 (T0) to 165.8 $\pm$ 15.4 mm^2 (T1) ($p > 0.05$) (Fig. 7).

In the eyes closed condition, the ellipse surface also was reduced in both groups. The reduction was significant in Group A from 168.5 $\pm$ 13.4 mm^2 to 153.2 $\pm$ 9.7 mm^2 ($p < 0.05$) and insignificant in Group B from 171.4 $\pm$ 22.0 mm^2 to 159.4 $\pm$ 16.9 mm^2 ($p > 0.05$) at T0 and T1, respectively (Fig. 8). Likewise, concerning the sway area, a significant modification was found only in Group A in both

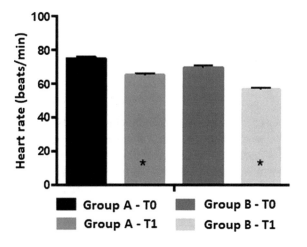

Fig. 1 Heart rate at protocol onset (T0) and mid-protocol fourth session (T1) in Group A, treated with neuromuscular manual therapy, and in Group B, treated with back massage therapy; *p < 0.05 *vs.* the corresponding heart rate in each patient group

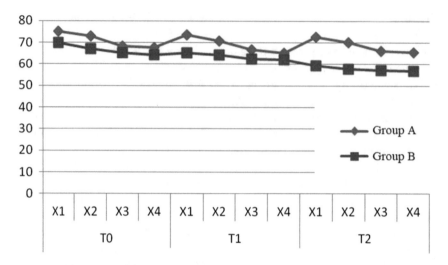

Fig. 2 Heart rate at protocol onset (T0), mid-protocol fourth session (T1), and the protocol-ending eighth session (T2) in Group A, treated with neuromuscular manual therapy, and in Group B, treated with back massage therapy. Each of these main sessions was subdivided into four time marks, i.e., onset (baseline), 5 min from onset, end, and 5 min after session end, marked as X1, X2, X3, and X4, respectively

Fig. 3 Breathing rate at protocol onset (T0) and mid-protocol fourth session (T1) in Group A, treated with neuromuscular manual therapy, and in Group B, treated with back massage therapy; *p < 0.05 *vs.* the corresponding breathing rate in each patient group

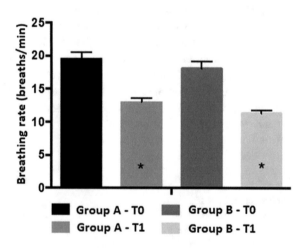

eyes open and closed conditions (Figs. 9 and 10). The sway length decreased significantly in both eyes open and closed conditions in Group A, treated with neuromuscular manual therapy, from 313.6 ± 19.7 mm^2 at T0 to 272.6 ± 14.3 mm^2 at T1 (p < 0.01) and from 313.3 ± 22.3 mm^2 at T0 to 292.1 ± 17.2 mm^2 at T1 (p < 0.05), respectively. In Group B, treated with back massage therapy, the respective decreases were insignificant, from 279.8 ± 25.4 mm^2 at T0 to 265.3 ± 18.9 mm^2 at

T1 in the eyes open (Fig. 9) and from 297.4 ± 23.2 mm^2 at T0 to 294.3 ± 20.5 mm^2 at T1 in the eyes closed (Fig. 10) conditions.

The patients of both groups achieved a significant improvement in pain perception in response to both types of therapies as assessed on the visual analog scale. The perception of pain decreased, on average, from 7.0 ± 2.8 points at T0 to 1.6 ± 0.8 points at T1 in Group A (Fig. 11) and from

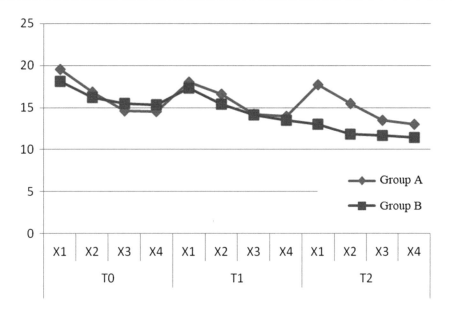

Fig. 4 Breathing rate at protocol onset (T0), mid-protocol fourth session (T1), and the protocol-ending eighth session (T2) in Group A, treated with neuromuscular manual therapy, and in Group B, treated with back massage therapy. Each of these main sessions was subdivided into four time marks, i.e., onset (baseline), 5 min from onset, end, and 5 min after session end, marked as X1, X2, X3, and X4, respectively

Fig. 5 Oxygen saturation (SpO$_2$) at protocol onset (T0) and mid-protocol fourth session (T1) in Group A, treated with neuromuscular manual therapy, and in Group B, treated with back massage therapy; *p < 0.05 *vs.* the corresponding oxygen saturation in each patient group

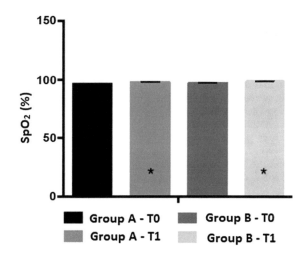

7.0 ± 3.1 points at T0 to 3.8 ± 1.6 points at T1 (Fig. 12).

4   Discussion

It is known that the autonomic effects of somatic stimulation depend on a particular organ and on its specific spinal afferent segmental signals. In anesthetized animals in which emotional factors

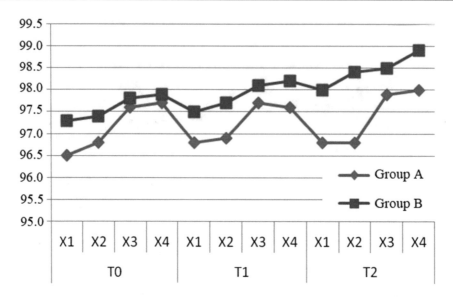

Fig. 6 Oxygen saturation at protocol onset (T0), mid-protocol fourth session (T1), and the protocol-ending eighth session (T2) in Group A, treated with neuromuscular manual therapy, and in Group B, treated with back massage therapy. Each of these main sessions was subdivided into four time marks, i.e., onset (baseline), 5 min from onset, end, and 5 min after session end, marked as X1, X2, X3, and X4, respectively

Fig. 7 Ellipse surface with extra rotation of feet at 30° in the eyes open condition at protocol onset (T0) and mid-protocol fourth session (T1) in Group A, treated with neuromuscular manual therapy, and in Group B, treated with back massage therapy; *p < 0.05 *vs.* the corresponding ellipse surface in Group A patients; ns, non-signficant

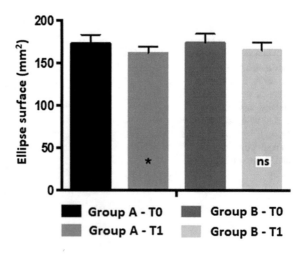

are eliminated, somatic afferent neural stimulation can regulate various visceral functions (Kimura et al. 1996). All structures receiving efferent fibers from the same spinal segment would be potentially exposed to excessive excitation or inhibition, which can give rise to a condition of self-sustained hyperactivity in the area of somatic afferents referred to as the facilitation state (Pettman 2007; McCracken and Turk 2002; Saggini et al. 1996; Vecchiet et al. 1991; Korr 1978; Johansson 1962; Korr et al. 1962).

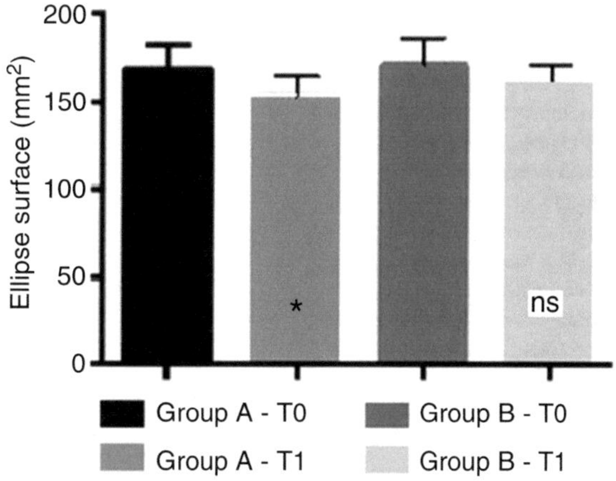

Fig. 8 Ellipse surface with extra rotation of feet at 30° in the eyes closed condition at protocol onset (T0) and mid-protocol fourth session (T1) in Group A, treated with neuromuscular manual therapy, and in Group B, treated with back massage therapy; *p < 0.05 *vs.* the corresponding ellipse surface in Group A patients; ns, non-significant

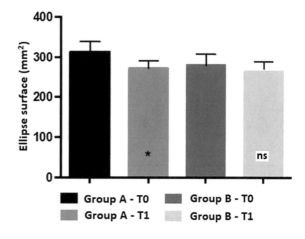

Fig. 9 The sway length with extra rotation of feet at 30° in the eyes open condition at protocol onset (T0) and mid-protocol fourth session (T1) in Group A, treated with neuromuscular manual therapy, and in Group B, treated with back massage therapy; *p < 0.05 *vs.* the corresponding ellipse surface in Group A patients; ns, non-significant

The findings of the present study confirmed the existence of somato-visceral pathways that can be activated through both types of somatic stimulation employed, i.e., neuromuscular manual therapy and classical back massage therapy. Some evidence suggests that the postganglionic sympathetic efferents are involved in the mediation of peripheral inflammatory responses through interaction with the primary afferent terminals (primary afferent nociceptors) (Michaelis et al. 2000; Vecchiet et al. 1999; Miao et al. 1996; Jänig 1996). Clinical evidence also indicates that the peripheral pathways of the sympathetic nervous system contribute to inflammation. For example, reflex sympathetic dystrophy manifests itself with pain, sympathetic hyperactivity, and inflammation of synovial joints (Kozin et al. 1976). Blocking the regional sympathetic activity with guanethidine or other sympatholytic agents may reduce the inflammatory state. Moreover, peripheral interaction between the primary afferent nociceptors and the sympathetic efferents may increase inflammation. There is a facilitatory interaction between the sympathetic efferents and the sensory afferents at the neuronal level and also in heat sensitive afferent

Fig. 10 The sway length with extra rotation of feet at 30° in the eyes closed condition at protocol onset (T0) and mid-protocol fourth session (T1) in Group A, treated with neuromuscular manual therapy, and in Group B, treated with back massage therapy; *$p < 0.05$ *vs.* the corresponding ellipse surface in Group A patients; non-significant

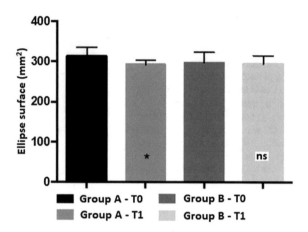

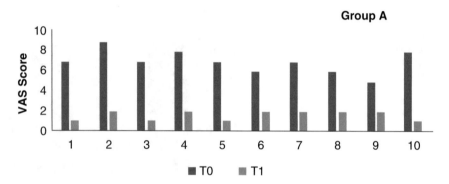

Fig. 11 Perception of pain on a visual analog scale (VAS) at protocol onset (T0) mid-protocol fourth session (T1) in Group A, treated with neuromuscular manual therapy

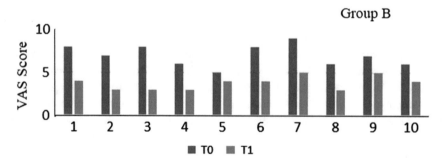

Fig. 12 Perception of pain on a visual analog scale (VAS) at protocol onset (T0) and mid-protocol fourth session (T1) in Group B, treated with back massage therapy

fibers. The activation of nociceptive afferents also increases the activity of postganglionic sympathetic efferents. Disrupting these interactions may provide prolonged anti-inflammatory effects in a variety of pathological states (Levine et al. 1986). Therefore, both treatment modalities employed in the present study were conducive to a reduction in spinal inflammation and pain perception, which was, in all likelihood, mediated by the autonomic effects.

Melzack (1999) and Kimura et al. (1996) have attempted to explain how the sensation is modulated at the spinal cord level. A sudden pain is interpreted as trauma that threatens the tissues; prolonged pain is interpreted centrally as a need for a longer rest that allows for the recovery from trauma. When pain persists beyond the natural duration of the stimulus or the pathology that produces it, it becomes chronic. This mechanism happens, for example, when myofascial issues are not adequately treated (Mendell 2014). It is reported that stimulation of specific points on the body surfaces during acupuncture therapy could effectively ameliorate the general and visceral pain perception, psychoneurotic disorders, as well as other ailments (Sato et al. 1986; Sato and Schmidt 1971). A local musculoskeletal dysfunction can possibly cause a continuous activation of local nociceptors, which initiates or sustains central sensitization. Thus, effective manual somatic stimulation in (sub)acute cases should be able to limit the time course of afferent barrage of noxious input to the central nervous system and thus prevent chronicity.

In addition, neuromuscular manual therapy, aimed at improving the motor control in symptomatic regions/joints, is likely to have a role in the prevention of chronic pain or dysfunction. Indeed, a sustained mismatch between the motor activity and the sensory feedback is able to serve as an ongoing source of nociception within the central nervous system. It is difficult to state specifically or even to generalize upon which autonomic component will dominate as the efferent path in these reflexes, because this depends on the individual organ, the site being stimulated, and the nature or mode of the stimulation (Sakai et al. 2007).

There is evidence that a mismatch between the motor activity and the sensory feedback can elicit pain and sensory perceptions in healthy pain-free volunteers (McCabe et al. 2005) and can exacerbate pain and sensory perceptions in patients with fibromyalgia (McCabe et al. 2007), suggesting a possible etiological role for sensomotor inconsistencies in the development of (chronic) pain. The role of the motor control system in the brain is to manage the relationship between motor commands and the sensory feedbacks (proprioception, vision). In case of an inaccurate execution of movements, due to motor deconditioning or joint tissue damage (and consequent altered proprioception), a mismatch between motor activity and sensory feedback is likely to occur. The motor control system may alert the individual of the abnormality in information processing by generating warning signals (i.e., pain or other sensory changes, like temperature change) (McCabe et al. 2005). Disrupting this aberrant circle with an afferent therapeutic stimulus in case of non-specific low back pain, may suffice to interact with dysfunction. The findings of the present study show that both neuromuscular manual therapy and spine massage facilitate the appearance of beneficial effects through such a mechanism.

Several studies have explored the neurophysiological basis of specific manual techniques in the cervical spine and upper limbs, utilizing the sympathetic nervous system function as a measure of response (Sterling et al. 2001; Vicenzino et al. 1994). Specific sympathetic responses, sudomotor function, cutaneous vasomotor changes, and cardiac and respiratory functions, have been reported after neuromuscular manual treatment (Chiu and Wright 1996). Muscles and fascia often become hypertonic in people with chronic widespread pain, which becomes a trigger point defined by local dysfunction (Schleip et al. 2005). The myofascial trigger points differ from the normal muscle tissue by a lower pH level, i.e., higher local acidity, increased levels of substance P, and the presence of gene-related calcitonin peptide, tumor necrosis factor-alpha, and interleukin-1beta; each having a role in increased pain sensitivity. Sensitized muscle nociceptors are more easily activated and may respond to normally innocuous and weak stimuli such as light pressure and muscle movement.

Therefore, it is recommended in neuromuscular manual therapy to start superficially with a soft-tissue mobilization, with well-tolerated strokes along the length of muscle fibers, referred to as 'stripping' (Benjamin and Tappan 2005), and to progress through deeper strokes that go perpendicular on the soft-tissue fibers. Neuromuscular manual therapy stimulate intrafascial mechanoreceptors, producing a change of the proprioceptive afferents directed to the central nervous system that, in turn, leads to a change in the regulation of the tone of the involved tissue (Granger 2011; Nijs et al. 2006).

The nervous system works across functional units of the motor system. There are several million of motor units in the body. Depending on the quality of the feedback mediated by sensory neurons, these units can be adjusted individually (Henatsch 1976). It is observed that muscle plasticity depends on the density of mechanoreceptors the muscle contains and it is not purely related to mechanical factors. In particular, the Ruffini receptors and so-called interstitial receptors can trigger changes in the autonomic nervous system function. Stimulation of these sensory endings can lead to changes in tone of motor units that are mechanically interrelated and connected with the fascial tissues that contain them; the effects highlighted by the present findings in Group B patients treated with back massage. The neuromuscular stimulation, acting on intra-fascial receptors, would activate smooth muscle cells embedded in the collagen fibers and nerves. It is thus probable that these cells, through the autonomic nervous system, can adjust a sort of pre-fascial tension, regardless of the muscle tone, which would operate as an accessory control system of posture and movement (Schleip 2003). Somatic stimulation through neuromuscular manual therapy, after a structural evaluation of the area in state of facilitation, can lead to both autonomic and postural improvements, like the evidence above outlined demonstrates.

In conclusion, autonomic variables seem suitable for the assessment of peripheral tissue stimulation, postural adjustments, and pain. The study suggests that neuromuscular manual treatment is capable of activating the central mechanisms responsible for pain control and for the modulation of autonomic functions and posture.

Conflicts of Interest The authors declare no conflicts of interest in relation to this article.

References

American Osteopathic Association (2010) Guidelines for Osteopathic Manipulative Treatment (OMT) for patients with low back pain. Task Force on the low back pain clinical practice guidelines. J Am Osteopath Assoc 116(8):536–549

Balagué F, Mannion AF, Pellisé F, Cedraschi C (2012) Non-specific low back pain. Lancet 379 (9814):482–491

Benjamin PJ, Tappan FM (2005) Tappan's handbook of healing massage techniques: classic, holistic and emerging methods, 4th edn. Pearson/Prentice-Hall, Saddle River

Cervero F, Connell LA (1984) Distribution of somatic and visceral primary afferent fibers within the thoracic spinal cord of the cat. J Comp Neurol 230(1):88–98

Chiu TW, Wright A (1996) To compare the effects of different rates of application of a cervical mobilisation technique on sympathetic outflow to the upper limb in normal patients. Man Ther 1(4):198–203

Cholewicki J, McGill SM (1996) Mechanical stability of the in vivo lumbar spine: implications for injury and chronic low back pain. Clin Biomech (Bristol, Avon) 11(1):1–15

Fryer G, Morris T, Gibbons P (2004) The relationship between palpation of thoracic paraspinal tissues and pressure sensitivity measured by a digital algometer. J Osteopath Med 7:64

Gori L, Firenzuoli F (2005) Posturology. Methodological problems and scientific evidence. Recenti Prog Med 96 (2):89–91. (Article in Italian)

Granger J (2011) Neuromuscular therapy manual. Wolters Kluver/Lippincott Williams & Wilkins, Baltimore

Henatsch HD (1976) Bauplan der peripheren und zentralen sensomotorischen Kontrollen. In Gauer OH, Kramer K, Jung R (eds) Physiologie des Mensche. Bd XIV: Sensomotorik. Urban & Schwarzenberg, München (Article in German)

Jänig W (1996) Neurobiology of visceral afferent neurons: neuroanatomy, functions, organ regulations and sensations. Biol Psychol 42(1–2):29–51

Johansson B (1962) Circulatory responses to stimulation of somatic afferents with special reference to depressor effects from muscle nerves. Acta Physiol Scand Suppl 198:1–91

Kapteyn TS, Bles W, Njiokiktjien CJ, Kodde L, Massen CH, Mol JM (1983) Standardization in platform stabilometry being a part of posturography. Agressologie 24(7):321–326

Katz JN (2006) Lumbar disc disorders and low-back pain: socioeconomic factors and consequences. J Bone Joint Surg Am 88(2):21–24

Kimura A, Sato A, Sato Y, Suzuki A (1996) Single electrical shock of a somatic afferent nerve elicits A- and C-reflex discharges in gastric vagal efferent nerves in anesthetized rats. Neurosci Lett 210(1):53–56

Koes BW, Van Tulder MW, Thomas S (2006) Diagnosis and treatment of low back pain. BMJ 332 (7555):1430–1434

Korr IM (1978) Sustained sympathicotonia as a factor in disease. In: Korr IM (ed) The neurobiologic mechanisms in manipulative therapy. Springer, Boston

Korr IM, Wright HM, Thomas PE (1962) Effects of experimental myofascial insults on cutaneous patterns of sympathetic activity in man. Acta Neuroveg 23:329–355

Kozin F, McCarty DJ, Sims J, Genant H (1976) The reflex sympathetic dystrophy syndrome: I. Clinical and histologic studies: evidence for bilaterality, response to corticosteroids and articular involvement. Am J Med 60(3):321–331

Levine JD, Dardick SJ, Roizen MF, Clyde H, Allan I (1986) Basbaum contribution of sensory afferents and sympathetic efferents to joint injury in experimental arthritis. J Neurosci 6(12):3423–3429

Licciardone JC (2008) The epidemiology and medical management of low back pain during ambulatory medical care visits in the United States. Osteopath Med Prim Care 2(1):11

McCabe CS, Haigh RC, Halligan PW, Blake DR (2005) Simulating sensory–motor incongruence in healthy volunteers: implications for a cortical model of pain. Rheumatology 44(4):509–516

McCabe CS, Cohen H, Blake DR (2007) Somaesthetic disturbances in fibromyalgia are exaggerated by sensory–motor conflict: implications for chronicity of the disease? Rheumatology 46(10):1587–1592

McCracken LM, Turk DC (2002) Behavioural and cognitive-behavioural treatment for chronic pain. Outcome, predictors of outcome, and treatment process. Spine 27(22):2564–2525

Melzack R (1999) From the gate to the neuromatrix. Pain 6:121–126

Mendell LM (2014) Constructing and deconstructing the gate theory of pain. Pain 15(2):210–216

Miao FJ, Jänig W, Green PG, Levine JD (1996) Inhibition of bradykinin-induced synovial plasma extravasation produced by intrathecal nicotine is mediated by the hypothalamopituitary adrenal axis. J Neurophysiol 76 (5):2813–2821

Michaelis M, Liu X, Jänig W (2000) Axotomized and intact muscle afferents but no skin afferents develop ongoing discharges of dorsal root ganglion origin after peripheral nerve lesion. J Neurosci 20(7):2742–2748

Nijs J, Meeus M, De Meirleir K (2006) Chronic musculoskeletal pain in chronic fatigue syndrome: recent developments and therapeutic implications. Man Ther 11(3):187–191

O'Sullivan PB, Twomey L, Allison GT (1997) Dysfunction of the neuro-muscular system in the presence of low back pain – implications for physical therapy management. J Man Manip Ther 5(1):20–26

Patterson MM, Wurster RD (2011) Somatic dysfunction, spinal facilitation, and viscerosomatic integration. In: Chila AG (ed) Foundations of osteopathic medicine, 3rd edn. Lippincott William & Wilkins, Philadelphia

Pengel L, Herbert R, Maher CG, Refshauge K (2003) Acute low back pain: systematic review of its prognosis. BMJ 327:323–327

Pettman E (2007) A history of manipulative therapy. J Man Manip Ther 15(3):165–174

Price DD, McGrath PA, Rafii A, Buckingham B (1983) The validation of visual analogue scales as ratio scale measures for chronic and experimental pain. Pain 17 (1):45–56

Saggini R, Ridi R (2002) Equilibrio corporeo. Martina Editore, Bologna

Saggini R, Giamberadino MA, Gatteschi L, Vecchiet L (1996) Myofascial pain syndrome of the peroneus longus: biomechanical approach. Clin J Pain 12:30–37

Sakai S, Hori E, Umeno K, Kitabayashi N, Ono T, Nishijo H (2007) Specific acupuncture sensation correlates with EEGs and autonomic changes in human subjects. Auton Neurosci 133(2):158–169

Sato A, Schmidt RF (1971) Spinal and supraspinal components of the reflex discharges into lumbar and thoracic white rami. J Physiol 212(3):839–850

Sato A, Sato Y, Schmidt RF (1986) Catecholamine secretion and adrenal nerve activity in response to movements of normal and inflamed knee joints in cats. J Physiol 375:611–624

Schleip R (2003) Fascial plasticity: a new neurobiological explanation: Part 1. J Bodyw Mov Ther 7(1):11–19

Schleip W, Klingler F, Horn L (2005) Active fascial contractility: fascia may be able to contract in a smooth muscle-like manner and thereby influence musculoskeletal dynamics. Med Hypotheses 65:273–277

Sterling M, Jull G, Wright A (2001) Cervical mobilisation: concurrent effects on pain, sympathetic nervous system activity and motor activity. Man Ther 6(2):72–81

Travell JG, Simons DG (1992) Myofascial pain and dysfunction: the trigger point manual, 2nd edn. Simons DG, Travell JG, Lois LS (eds). Williams & Wilkins, Baltimore

Vecchiet L, Giamberardino MA, Saggini R (1991) Myofascial pain syndromes: clinical and pathophysiological aspects. Clin J Pain Suppl 7:S16–S22

Vecchiet L, Vecchiet J, Giamberardino MA (1999) Referred muscle pain: clinical and pathophysiologics aspects. Curr Rev Pain 3:489–498

Vicenzino B, Collins D, Wright T (1994) Sudomotor changes induced by neural mobilisation techniques in asymptomatic patients. J Man Manip Ther 2(2):66–74

Advs Exp. Medicine, Biology - Neuroscience and Respiration (2018) 39: 111–115
DOI 10.1007/5584_2018_158
© Springer International Publishing AG 2018
Published online: 13 Feb 2018

Robot-Assisted Body-Weight-Supported Treadmill Training in Gait Impairment in Multiple Sclerosis Patients: A Pilot Study

Marek Łyp, Iwona Stanisławska, Bożena Witek, Ewelina Olszewska-Żaczek, Małgorzata Czarny-Działak, and Ryszard Kaczor

Abstract

This study deals with the use of a robot-assisted body-weight-supported treadmill training in multiple sclerosis (MS) patients with gait dysfunction. Twenty MS patients (10 men and 10 women) of the mean of 46.3 ± 8.5 years were assigned to a six-week-long training period with the use of robot-assisted treadmill training of increasing intensity of the Lokomat type. The outcome measure consisted of the difference in motion-dependent torque of lower extremity joint muscles after training compared with baseline before training. We found that the training uniformly and significantly augmented the torque of both extensors and flexors of the hip and knee joints. The muscle power in the lower limbs of SM patients was improved, leading to corrective changes of disordered walking movements, which enabled the patients to walk with less effort and less assistance of care givers. The torque augmentation could have its role in affecting the function of the lower extremity muscle groups during walking. The results of this pilot study suggest that the robot-assisted body-weight-supported treadmill training may be a potential adjunct measure in the rehabilitation paradigm of 'gait reeducation' in peripheral neuropathies.

Keywords

Gait · Joint function · Lower extremity · Multiple sclerosis · Muscle strength · Robot-assisted muscle actuator · Torque · Treadmill training

1 Introduction

Multiple sclerosis (SM) is a chronic, demyelinating central nervous system disease. The severity and the rate of progression of MS is variable, with the disease course being often dependent on the form, degree, and location of lesions. Maintaining physical activity and muscle rehabilitation remain the essential part of a treatment plan to extend the patient everyday functioning. To this end, recent technological advances in the form of a robot-

M. Łyp, I. Stanisławska (✉), E. Olszewska-Żaczek, and R. Kaczor
Department of Physiotherapy, College of Rehabilitation, Warsaw, Poland
e-mail: iwona.stanislawska@wsr.edu.pl

B. Witek
Department of Animal Physiology, Institute of Biology, The Jan Kochanowski University in Kielce, Kielce, Poland

M. Czarny-Działak
Faculty of Medicine and Health Sciences, The Jan Kochanowski University in Kielce, Kielce, Poland

assisted patient-tailored therapy, which exploits neuroplasticity and neuromuscular functional reserve during rehabilitation, are increasingly used. That notably concerns the methods that are expected to restore the physiologically symmetrical gait-orthosis, most often defective in neuropathies (Kumru et al. 2016a, b; Swinnen et al. 2012). Robot-assisted support can be adjusted to shape therapy intensity, which appears therapeutically advantageous. Therefore, the purpose of the present study was to examine the effect of a robot-assisted body-weight-supported treadmill training on the walking ability of SM patients with impaired gait. The outcome measures consisted of changes in motion-dependent joint torque of lower extremity muscles.

2   Methods

The study was approved by the Ethics Committee of the College of Rehabilitation in Warsaw, Poland (permit no. 53/2015) and was conducted according to the principles of the Declaration of Helsinki for Human Research. The participants were informed about the research purpose and provided informed consent to participate in the study. The study was conducted in at the National Center for Multiple Sclerosis patients in the town of Dabek, Poland, between November 2015 and February 2016. There were 20 SM patients (10 women and 10 men) of the mean age of 46.3 ± 8.5 years, diagnosed according to the 2010 Revision of the McDonald criteria (Polman et al. 2010). The patients were free of comorbidities. Seven patients were diagnosed with relapsing-remitting SM, with no relapse during the preceding 6 months, and 13 patients with primarily progressive SM.

The robot-assisted body-weight-supported treadmill training was spread over a period of 6 weeks. Each patient underwent two sessions of training a week, always on Tuesdays and Thursdays, which makes up 12 sessions in total. The first initiating training lasted for 15 min, the second lasted for 25 min, and all following training sessions lasted for 35 min. Treadmill walking

was without an incline but with the load equal to half of the patient's body weight. The basic speed was 1 km/h and was steadily increasing over the session time up to 1.8 km/h.

We used the Lokomat System (Hacoma; Volketswil, Switzerland) for rehabilitative training of the extension and flexion muscles of the hip and knee joints in MS patients. The setup consisted of an instrumented treadmill, computer-controlled dynamic body weight relief with four orthoses supporting the lower limbs. The patients' legs were attached to the Lokomat using instrumented shank and thigh cuffs, as well as foot lifters to help foot clearance in the swing phase, equipped with the attached motion tracking markers and muscle force sensors. The motion capture system tracked the interaction of each limb fragments. The system was controlled by the proprietary computer software, with the intuitive user interface and the visual feedback control of rehabilitation progress, available for both therapist and patient. The system was thus highly motivational as it enabled the patient to follow the effect of training on augmentative muscle performance. The Lokomat assistive setup picks up the EMG signals before the muscle contracts when the trainee performs the task movement on the treadmill. The signals are then converted into estimated joint torque. The lower extremity joint torques were quantified in Newton-meter (Nm) units and were used as a surrogate of the strength of the hip and knee muscle groups. The torque was quantified using the Lokomat system's software prior to onset of training and directly after the end of each session.

Data were presented as the mean $\pm$ SE difference (Δ) between the baseline level of torque of hip and knee joint muscles, assessed before and after rehabilitative treatment. Ninety five percent confidence intervals (95%CI) for the means were calculated. Data distributions was tested using the Kolmogorov-Smirnov test. The pre-post differences in the joint torque were compared with a two-tailed paired t-test. A p-value < 0.05 defined the statistical significance of differences. A commercial IBM SPSS v24.0 statistical package (SPSS Corp; Armonk, NY) was used for all data evaluations.

3   Results and Discussion

The quantification of torque changes of the hip and knee joint muscles in MS patients after a robot-assisted body-weight-supported treadmill training of Lokomat type is depicted in Table 1. This table collates mean differences in joint muscle torque achieved after the training against the baseline unsupported level of the corresponding muscle torque. The rehabilitative training uniformly augmented the torque of both extensors and flexors of both joints; the augmentations were significant and were closely symmetrical for the corresponding joints of both lower limbs.

In this study we used motion-dependent joint torque in a swing phase of legs as a surrogate of the mechanical power of respective muscle groups. The aim was to assess the potential benefits of assisted rehabilitative training for gait disturbance in MS patients. We found an augmentation in motion-dependent torques of the muscles controlling the hip and knee joints' walking movements. Torque augmentation was apparently caused by the external force of robotic-enhancement acting on lower extremity joints to rapidly alternate between flexion and extension during the swing phase of gait. The torque augmentation may have its role in affecting the function of the lower extremity muscle groups during walking. The quadriceps muscle group is the prime knee extensor and higher torque values may reflect an improvement in the body posture and stabilization (Hart et al. 1984). It might thus be surmised that the general augmentation in the muscle torque we observed could serve to counteract the disease-dependent distortion of joint movements in MS to better control and synchronize the gait direction of lower limbs. These findings may help better understand the function of the lower extremity muscle groups during 'gait reeducation' in peripheral neuropathies.

The concept of a torque of joint muscles relates to the basic force that comes from a group of muscles to overcome the joint's turning resistance (Hoy et al. 1990). That is a force that is required for torsion or rotation of a joint during the act of walking, which involves flexion and extension of respective muscles. Thus, two opposite acting torques are harmonized to stabilize the posture and direction of walking. The moment of force causing joint rotation derives from a complex mix of interactive components including muscle mass and fiber length, muscle groups involved in motion, limb length, and the type of disease-driven function distortion. These components are reflected in the final assessment of muscle torque, which is further complicated by a three dimensional character of human joint movement (Lieber and Shoemaker 1992; McClearn 1985). A detailed componential evaluation of the lower extremity muscle torque was beyond the scope of this pilot study as it requires alternative experimental designs. We believe, however, that the effect on the hip and knee joint torque of robot-assisted treadmill walking we have herein reported points to the usefulness of this kind of training in correcting gait disturbance in SM patients. An optimal way to correct impaired gait in neurological pathologies remains uncertain, but such correction is of essential importance from the bio-psychological standpoint and the patient's perception of quality of life. Therefore, potential benefits of a robot-assisted body-weight-supported treadmill training warrant further exploration.

Several other studies have examined the effects of robot-assisted motion actuators on lower extremity joint torques. The Lokomat robotic-enhanced muscle-joint training effectively improves gait in SM patients (Schwartz et al. 2012), children with cerebral palsy (Wallard et al. 2017), and in spinal cord injury (Nam et al. 2017; Kumru et al. 2016a, b). Such training system also improves the cognitive and topographic abilities of patients after stroke and improves the motor and cognitive abilities of patients with vascular dementia (Calabrò et al. 2015). The added value of the system is the possibility of adjustment of training parameters in a feedback manner, which distinctly helps obtain good therapeutic results (van Kammen et al. 2016; Aurich et al. 2015). On the other side, there are reports contradicting any efficacy of the robot-assisted systems greater and beyond more classical rehabilitation methods, concerning also the lack of benefits in gait disorders or quality of

Table 1 Torque increments (Δ Nm) in the extension and flexion muscles of hip and knee joints after a robot-assisted body-weight-supported treadmill training

	Δ Torque	95%CI	p
Extensors – left hip	5.44 ± 1.28	8.11–2.76	<0.001
Flexors – left hip	4.89 ± 1.46	7.95–1.82	<0.003
Extensors – left knee	5.97 ± 1.29	8.68–3.26	<0.001
Flexors – left knee	6.36 ± 1.60	9.70–3.01	<0.001
Extensor – right hip	6.55 ± 1.76	10.22–2.87	<0.001
Flexors – right hip	3.68 ± 1.64	7.12–0.24	<0.037
Extensors – right knee	6.11 ± 1.35	8.91–3.30	<0.001
Flexors – right knee	7.06 ± 1.95	11.14–2.97	<0.001

Data are means ±SE of difference (Δ) in the joint torque after training compared with baseline; 95% CI, 95% confidence interval of the mean; Nm, Newton-meter

daily-life (Dierick et al. 2017; Vaney et al. 2012; Wier et al. 2011; Neckel et al. 2008). The robot-assisted training systems remain an interesting developmental component of a comprehensive physical rehabilitation program in patients with neurological morbidities (Gandara-Sambade et al. 2017; Nam et al. 2017; van Kammen et al. 2017; Gor-García-Fogeda et al. 2016; Morawietz and Moffat 2013; Beer et al. 2008).

We conclude that the mission of the robot-assisted muscle-joint actuator is to provide safe and optimized inpatient rehabilitative training, with the capability to monitor rehabilitation progress in patients with chronic neurological disorders. We believe this mission was fulfilled by the Lokomat system used in the present study in a six-week-long training program. The muscle power in the lower limbs of SM patients was apparently improved in a closely symmetrical manner, leading to corrective changes of disordered gait pattern, which enabled the patients to walk with less effort and less relying on external walk-helping instruments or care givers' assistance. The interactive feedback-controlled robot-assisted body-weight-supported treadmill training constitutes a bio-psychological paradigm of rehabilitation treatment. The paradigm might be suitable for patients, who despite the use of various pharmacological and physical treatment modalities, continue to display progressive physical difficulties in negotiating daily-life activities, notably those linked to gait dysfunction. In addition, this mode of rehabilitation appears user-friendly for both therapists and patients. Thus, despite the lack of clearly verifiable superiority over classical therapeutic modes, but also of any negative sequelae, the robot-assisted treadmill training may be a potential adjunct measure in rehabilitation of neurological patients.

Conflicts of Interest The authors declare no conflicts of interest in relation to this article.

References

Aurich T, Warken B, Graser JV, Ulrich T, Borggraefe I, Heinen F, Meyer-Heim A, van Hedel HJ, Schroeder AS (2015) Practical recommendations for robot-assisted treadmill therapy (Lokomat) in children with cerebral palsy: indications, goal setting, and clinical implementation within the WHO-ICF framework. Neuropediatrics 46:248–260

Beer S, Aschbacher B, Manoglou D, Gamper E, Kool J, Kesselring J (2008) Robot-assisted gait training in multiple sclerosis: a pilot randomized trial. Mult Scler 14(2):231–236

Calabrò RS, De Luca R, Leo A, Balletta T, Marra A, Bramanti P (2015) Lokomat training in vascular dementia: motor improvement and beyond! Aging Clin Exp Res 27:935–937

Dierick F, Dehas M, Isambert JL, Injeyan S, Bouché AF, Bleyenheuft Y, Portnoy S (2017) Hemorrhagic versus ischemic stroke: who can best benefit from blended conventional physiotherapy with robotic-assisted gait therapy? PLoS One 12(6):e0178636

Gandara-Sambade T, Fernandez-Pereira M, Rodriguez-Sotillo A (2017) Robotic systems for gait re-education in cases of spinal cord injury: a systematic review. Rev Neurol 64(5):205–213. (Article in Spanish)

Gor-García-Fogeda MD, Cano de la Cuerda R, Carratalá Tejada M, Alguacil-Diego IM, Molina-Rueda F (2016) Observational gait assessments in people with neurological disorders: a systematic review. Arch Phys Med Rehabil 97(1):131–140

Hart DL, Stobbe TJ, Till CV, Plummer RW (1984) Effect of muscle stabilization on quadriceps femoris torque. Phys Ther 64(9):1375–1380

Hoy MG, Zajac FE, Gordon ME (1990) A musculoskeletal model of the human lower extremity: the effect of muscle, tendon, and moment arm on the moment-angle relationship of musculotendon actuators at the hip, knee, and ankle. J Biomech 23:157–169

Kumru H, Benito-Penalva J, Valls-Sole J, Murillo N, Tormos JM, Flores C, Vidal J (2016a) Placebo-controlled study of rTMS combined with Lokomat® gait training for treatment in subjects with motor incomplete spinal cord injury. Exp Brain Res 234(12):3447–3455

Kumru H, Murillo N, Benito-Penalva J, Tormos JM, Vidal J (2016b) Transcranial direct current stimulation is not effective in the motor strength and gait recovery following motor incomplete spinal cord injury during Lokomat® gait training. Neurosci Lett 620:143–147

Lieber RL, Shoemaker SD (1992) Muscle, joint, and tendon contributions to the torque profile of frog hip joint. Am J Phys 263:R586–R590

McClearn D (1985) Anatomy of raccoon (Procyon lotor) and caoti (Nasua narica and N. nasua) forearm and leg muscles: relations between fiber length, moment-arm length, and joint excursion. J Morphol 183:87–115

Morawietz C, Moffat F (2013) Effects of locomotor training after incomplete spinal cord injury: a systematic review. Arch Phys Med Rehabil 94(11):2297–2308

Nam KY, Kim HJ, Kwon BS, Park JW, Lee HJ, Yoo A (2017) Robot-assisted gait training (Lokomat) improves walking function and activity in people with spinal cord injury: a systematic review. J Neuroeng Rehabil 14:24

Neckel ND, Blonien N, Nichols D, Hidler J (2008) Abnormal joint torque patterns exhibited by chronic stroke subjects while walking with a prescribed physiological gait pattern. J Neuroeng Rehabil 5:19

Polman CH et al (2010) Diagnostic criteria for multiple sclerosis: 2010 revisions to the McDonald criteria. Ann Neurol 69(2):292–302

Schwartz I, Sajin A, Moreh E, Fisher I, Neeb M, Forest A, Vaknin-Dembinsky A, Karusis D, Meiner Z (2012) Robot-assisted gait training in multiple sclerosis patients: a randomized trial. Mult Scler 18:881–890

Swinnen E, Beckwée D, Pinte D, Meeusen R, Baeyens JP, Kerckhofs E (2012) Treadmill training in multiple sclerosis: can body weight support or robot assistance provide added value? A systematic review. Mult Scler Int 2012:240274

van Kammen K, Boonstra AM, van der Woude LH, Reinders-Messelink HA, den Otter R (2016) The combined effects of guidance force, bodyweight support and gait speed on muscle activity during able-bodied walking in the Lokomat. Clin Biomech (Bristol, Avon) 36:65–73

van Kammen K, Boonstra AM, van der Woude LH, Reinders-Messelink HA, den Otter R (2017) Differences in muscle activity and temporal step parameters between Lokomat guided walking and treadmill walking in post-stroke hemiparetic patients and healthy walkers. J Neuroeng Rehabil 14(1):32

Vaney C, Gattlen B, Lugon-Moulin V, Meichtry A, Hausammann R, Foinant D, Anchisi-Bellwald AM, Palaci C, Hilfiker R (2012) Robotic-assisted step training (Lokomat) not superior to equal intensity of overground rehabilitation in patients with multiple sclerosis. Neurorehabil Neural Repair 26(3):212–221

Wallard L, Dietrich G, Kerlirzin Y, Bredin J (2017) Robotic-assisted gait training improves walking abilities in diplegic children with cerebral palsy. Eur J Paediatr Neurol 21:557–564

Wier LM, Hatcher MS, Triche EW, Lo AC (2011) Effect of robot-assisted versus conventional body-weight-supported treadmill training on quality of life for people with multiple sclerosis. J Rehabil Res Dev 48 (4):483–492

Advs Exp. Medicine, Biology - Neuroscience and Respiration (2018) 39: 117–118
https://doi.org/10.1007/978-3-319-89665-6

Index